A Handbook of Population Ecology

A Handbook of Population Ecology

H.K. Pathak

CYBER TECH PUBLICATIONS
4264/3, Ansari Road, Daryaganj, New Delhi-110002 (India)
Ph.: 011-23244078. Fax: 011-23280028
E-Mail: cyberpublicationsdelhi@yahoo.com

A HANDBOOK OF POPULATION ECOLOGY

H.K. Pathak

First Edition 2009

Published by :

G.S. Rawat for **Cyber Tech Publications**
4264/3, Ansari Road, Daryaganj, New Delhi-110002 (India)
Ph.: 011-23244078. Fax: 011-23280028
E-Mail: cyberpublicationsdelhi@yahoo.com

Printed at : **Amit Enterprises**, Maujpur, Delhi, (9810878184)

Preface

Practical ecological problems such as preservation of threatened species, design of nature reserves, planned harvest of game animals, management of fisheries, and evaluation of human impacts on natural systems are addressed with quantitative tools, such as models. A model is a mathematical representation of a natural process. Many biologists now use models implemented as computer software to approach the quantitative aspects of these practical problems.

In addition to their practical use, such models are excellent tools for developing a deeper understanding of how nature works. In an introduction to population ecology, most undergraduate students consider learning the mathematics required by traditional texts to be an unnecessary hindrance. The aim of this book is to teach quantitative methods that are necessary to develop a basic understanding and intuition about ecological processes, without intimidating or discouraging students who do not have extensive mathematical backgrounds. Even students who are intimidated by mathematical equations are not usually afraid of using computers. I hope that my integration of software that implements mathematical models in population ecology with an undergraduate textbook will make these models accessible to undergraduates in biological and environmental sciences.

—Author

CONTENTS

Preface

1. Introduction 1
2. Population Trends in Recent Decades 43
3. Behavioral and Social Sciences 75
4. Health Services Delivery 92
5. The Biomedical Ethics 111
6. Research on Aging 121
7. Human Cloning: Pros and Cons 134
8. Human Cloning: An Overview 152
9. Historical Aspects of Cloning 167
10. Scientific Background for Population Ecology 195
11. The Ethics of Cloning-to-Produce-Children 205
12. The Ethics of Cloning-for-Biomedical-Research 240

Bibliography 269

Index 271

1

Introduction

Science offers the best hope to improve the older person's quality of life. Research that is directed and supported properly can provide the means to reduce disability and dependence in old age, and can decrease the burdens on a health care system strained to its limits.

This is a time for celebration but a time as well for deep concern. It is a time for celebration because gains in life expectancy have resulted in an estimated 33 million Americans 65 years old and over, the majority of whom are vigorous and active well into advanced old age. It is a time for concern because, for a growing number of older people and their families, these added years of life often are burdened by disability, dementia, and the loss of independence.

In testimony before the Committee on a National Research Agenda on Aging, Joseph A. Califano, former Secretary of Health, Education, and Welfare, said:

> *The aging of America will challenge all our political, retirement, and social service systems. As never before, it will test our commitment to decent human values. Nowhere is the aging of America freighted with more risk and opportunitu than in the area of health care.*

The urgent need to respond to this risk is underscored by census figures showing that the subgroup of older persons most vulnerable to disability and dementia, those 85 years of age and over, is growing six times faster than the rest of the population, and by evidence indicating that the cost of caring for disabled older persons will more than double in the coming decades unless the causes of disability can be identified and controlled (Schneider and Guralnick, 1990).

The following themes of risk and opportunity have guided the preparation of this report, and have justified support for the study of aging through scholarship and research:

1. the need to control acute and chronic illness in old age and to achieve a major reduction in disability and suffering, thus not merely extending life but enhancing it; and
2. the new opportunity offered by science to improve the quality of life for older persons through analysis and discovery of the relationships among basic biological phenomena, disease, economic and social deprivation, disability, and dependence.

Building on two reports by the National Institute on Aging (NIA), *Our Future Selves* (1978), and *Toward an Independent Old Age: A National Plan for Research on Aging* (1982), the present agenda aims to identify priority research on aging in basic biomedical science, clinical science, and behavioral and social studies as they relate to health, health services delivery, and biomedical ethics. To this end, the Institute of Medicine (IOM) convened a committee of 18 leaders in national health care, and charged them to:

- identify for the next two decades priorities in research on aging that will contribute to the quality of life and maximize the independence of older persons;
- stimulate research on the fundamental processes of aging and the prevention of disability and disease, and further the application of current knowledge to the treatment of age-related disorders;
- alert scientists to promising but neglected areas of research on aging;
- provide the stimulus that will encourage new investigators to undertake the study of aging, and continue to engage the interest of established researchers in this field;
- estimate the resources required to fulfill the scientific needs for gerontological and geriatric research;

- guide federal and nonfederal funding agencies in supporting this research, and identify collaborative and nonduplicative funding patterns for government, industry, and private sources of support; and
- bring the importance of research on aging to the attention of government, private institutions, industry, the scientific community, and the public at large.

During the project the committee became convinced of the importance of ethical issues in clinical care and research on aging, and added this topic to the agenda.

Four liaison teams aided the committee: basic biomedical research, clinical research, social and behavioral research, and health services delivery research. Each of the teams was comprised of experts in their respective fields. Recommendations on biomedical ethics were provided by two national authorities in this field.

The reports of the liaison teams served as a major resource for the committee's deliberations. Guidance also was provided by five senior advisors who have made major contributions to the field of health care. Lastly, advice from experts in private foundations, government, industry, national organizations, and scientific institutions was obtained. The report, although assisted by information provided by more than 150 authorities, is the product of the committee's critical analysis of information received and represents the independent decisions of the committee.

Committee members and liaison team leaders developed criteria to identify priority research topics Although different fields of study emphasized different criteria, in general, priority research areas were selected for their potential to:

- contribute to the understanding of the basic mechanisms of aging;
- address problems of disability and functional impairment in older persons:
- increase knowledge both of the interaction between disease and aging and of age-related diseases;
- lower morbidity and mortality rates among older persons;

- be implemented in a timely and feasible manner, with short-term goals achievable in 5 years or less and long-term goals achievable in 5 to 20 years;
- lead to cost reduction in health care;
- increase knowledge of behavioral and social factors in health and disease, and help older people maintain social as well as biological health;
- improve pharmacologic treatments of patients; and
- attend to relatively neglected areas of investigation.

The 1982 *Toward an Independent Old Age: A National Plan for Research on Aging* lists more than 350 research possibilities; the present committee's agenda is restricted to 15 areas of priority research. As conceived, the research priorities listed in this report represent the conceptual issues that underlie the questions and specific propositions of research proposals.

The committee notes that in developing research priorities all five of the research areas were felt to be of first-rank importance, and the committee emphasizes that support should extend across all five of the fields of age-related research described in this report. Within each field of research, priorities are ranked according to their importance to the understanding of aging and their potential to improve the quality of life for older persons.

A national research agenda on aging has a broad multidisciplinary view, but it must be selective. The charge by the IOM dictated an agenda with the most promise for substantial impact on the individual during the next two decades—that is, research on the health and functioning of older individuals.

BACKGROUND

Support for Training and Research in Geriatrics and Gerontology

At the heart of health care for the older population are the educational programs that train the professionals who can respond to the health problems of elderly persons and make the scientific discoveries that will improve the quality of life of the later years. A 1980 study predicted that by the year 1990 training would be needed to produce 7,000 to 10,300 geriatricians to provide care for older persons having the most complex geriatric problems—those 75 years of age and over (Kane et al., 1980).

Programs for training have not met this goal, as shown by the finding that by the mid-1990s there will be only 5,000 certified geriatricians in this country (Reuben et al., 1990). To train an adequate number of geriatricians, Kane et al. (1980) estimated a full-time geriatric faculty of 1,230 to supervise hospital training programs in internal medicine or family practice, plus 370 to staff medical schools, and 450 for geropsychiatry, totaling about 2,100 faculty members and comprising both academic clinicians and researchers.

A 1987 IOM study, commenting on the small number of geriatric training programs (under 5 percent of all postgraduate medical programs), found that these programs graduated about 100 fellows per year and recommended that to meet needs for biomedical faculty by the year 2000 the number of graduates per year should be raised to at least 200 (IOM, 1987). Another study found training programs far short of goals for the year 2000 to develop nonbiomedical (mainly behavioral and social scientists) faculty in aging (National Institute on Aging, 1987). The serious shortage of faculty and researchers in age-related fields requires prompt attention by organizations that have traditionally supported this training (National Institute on Aging; Department of Veterans Affairs; National Institute of Mental Health).

Although training programs on aging, initially supported by the National Institute of Mental Health and the National Institute of Child Health and Human Development, began in the 1950s and 1960s, the establishment of the National Institute on Aging, the development of research and training programs in aging by the Department of Veterans Affairs, and the growth of scores of university programs in this country have taken place for the most part since 1970. Many discoveries in aging arising from this national effort—from molecular biology to behavioral and social science—await application to health problems and to improvement of function in older people.

Support is needed to continue these advances. The 1990 budget of the National Institutes of Health (NIH) devoted about $442 million, or about 6 percent of all funds, to the study of aging, with about $239 million coming from the NIA (including funds for research grants, centers, education, and demonstration projects) and an additional $203 million for research on aging coming from other parts of the NIH (Office of Planning, Technology, Information, and Evaluation, NIA). In 1990 other federal departments and agencies provided additional funds of about $144 million for research on aging. Foundations added $15 million in support.

The National Pharmaceutical Association lists $3.6 billion for research and development on drugs used to treat diseases occurring in older patients, but it is difficult to know how much of this money was designated for age-related research as such.

An increase in funding for research on aging could produce rapid advances in the ability to respond to the problems of many older persons. For example, the neurosciences are poised to make major discoveries in Alzheimer's disease, but for every 100 dollars spent on care for the victims of this disease, under 0.5 percent, or less than 50 cents, goes to research that might lead to the control of Alzheimer's disease (Advisory Panel on Alzheimer's Disease, 1989).

A recent article underscored the effect of restricted funds on research, noting that at the NIA the funding rate of approved traditional investigator- initiated research projects (R01s) for fiscal year 1989 was 16.9 percent, or about one in six (Movsesian, 1990). This percentage and that of all funded approved research project grants on aging (24 percent) are slightly lowered due to reapplications.

Low funding rates for research grants diminish opportunities for developing careers in age-related studies and may contribute to undersubscription to training programs in geriatrics (IOM, 1987). In 1990 support for research on aging from all sources (NIH + Department of Veterans Affairs + other federal departments and agencies + foundations), described further in Chapter 7, amounted to an estimated $601 million, or less than 0.5 percent of the estimated 1987 (the most recent year for which figures are available) $162 billion cost of care for disability and illness in older patients (Waldo et al., 1989). This small national commitment to research in aging is a wasteful strategy in light of the potential contribution of research to improve the status of older persons and to reduce the enormous and expanding costs of their care.

A growing sense of urgency is accompanied by the recognition that the time is right to bridge the gap between the needs of an aging society and the scientific knowledge base. The nation now needs a comprehensive plan for research on aging. By identifying research priorities and the resources to support research on aging, it is the committee's intent to respond to that need.

The Aging Society: Current Costs for Health Care of Older Persons

Most older persons are vigorous and have lives of good quality, and more than 50 percent of those 80 years and over are independent in self-care. Although mortality rates have fallen in the older generation, with increasing life expectancy, chronic diseases have become a major cause of death and disability. For instance, Rice reported that 40 percent of persons 75 and over had two or more chronic illnesses and that the proportion of the aged population reporting multiple chronic conditions had risen in recent years (Rice, 1989). Of the 10 leading causes of death among older persons, only 2—heart disease and diabetes mellitus—are listed among the 10 leading chronic geriatric conditions (National Center for Health Statistics, 1989). Most of the leading chronic conditions involve disability and prolonged decline. Disability here is defined as any restriction on or impairment in performing an activity in the manner or within the range considered normal for a human being (World Health Organization, 1982).

Chronic conditions contributing to disability include arthritis, heart disease, strokes, disorders of vision and hearing, nutritional deficiencies, and oral-dental problems (in up to 40 percent of older persons; Baum, 1988). Dementia (especially Alzheimer's disease), not listed as a leading cause of death or disability in older persons, nevertheless for those 75 years and older is a major contributor to disability, nursing home placement, and, indirectly, to mortality rates (Weiler, 1987).

Fewer than 7 percent of persons aged 65 to 74 require assistance with dressing or bathing. At 75 to 84, these activities require help in 7 percent to 14 percent of older persons, and at 85 and beyond, up to 30 percent will need help with bathing (National Center for Health Statistics, 1987). Nursing home placement is needed for those who lack social supports and are most limited in ability to care for themselves.

One half of nursing home residents require assistance with five or more activities of daily living (ADLs: bathing, dressing, grooming, transferring from bed to chair, going to the bathroom, being continent, feeding oneself), but for every one of the 1.4 million in skilled nursing facilities, two equally impaired older adults are receiving care at home. The cost of health care for older persons is great and growing rapidly. In 1987 the cost to provide health care for elderly patients was $162 billion or $5,360 per capita, compared with $45.2 billion ($1,856 per capita) a decade earlier (Waldo et al., 1989). Medicare costs alone were almost $75 billion for the care of disability and illness among old people in 1987, and they may rise to $145 billion per year (in 1987 dollars) by the year 2020 (Schneider and Guralnick, 1990).

Sick older people are burdened by accelerating costs for their own care. In 1980 the average yearly expenditure by older persons for health care was $966, and in 1988 it was $2,394. Costs for medical care borne by older people have more than doubled in the past decade and may triple in the coming decade (Select Committee on Aging, 1990).

The Potential of Research on Aging

The description of the compression of morbidity in old age (Fries, 1980), with the older person collapsing suddenly after many years of vigor, has stimulated a closer study of the relationship between morbidity and mortality in old age. In 1986 a 65-year-old person could look forward to 17 more years of life and an 85-year-old to 6 more years. By contrast, *active* life expectancy, which implies ability to care for oneself and be independent, was only 10 years for 65 to 70-year-olds, and for 85-year-olds it was only 2.9 years (Katz et al., 1983).

Diminishing mortality rates thus may swell the ranks of sick and debilitated elderly persons, and the number of years of dependent old age may increase, with ever greater outlays for health and medical care (Gruenberg, 1977; Schneider and

Guralnick, 1990). Like Tithonus, to whom the god Zeus gave eternal life without eternal youth (Graves, 1955), long life is a punishment if the quality of the later years is poor.

A model developed by Manton describes the potential of intervention to improve the quality of life for older persons. If medical care fails to diminish the incidence of disease among older persons, therapy that reduces lethal complications of disease may increase life expectancy in this group, but at the high cost of prolonging chronic illness and associated disability. Therefore, basic research should aim to promote interventions that decrease mortality by diminishing the incidence, severity, and rate of progression of disease.

This approach will then improve *active* life expectancy as well as life expectancy, with slowing of the rate of "aging," and with increased independence among older individuals (Manton and Soldo, 1985; Soldo and Manton, 1988; Fries, 1990). An example of this kind of intervention is seen in the recent advances in the prevention and treatment of the chronic degenerative disorder osteoporosis, and related hip fractures in older women.

The power of research to reduce health costs dramatically usually is underestimated. The lifetime cost to maintain two Rh brain-damaged children or two children severely crippled by poliomyelitis (and there were thousands of these children in 1960) was greater than all of the money spent on the research that virtually eliminated these conditions (J. Krevans, University of California, San Francisco, personal communication, 1990). Postponing by one month the onset of severe disability that leads to nursing home placement for older persons could lead to savings of $3 billion per year, minus the cost of providing care for these patients at home.

This is based on 1987 figures (the most recent available) for nursing home costs of $32.8 billion per year for those 65 years of age and over (Waldo et al., 1989), adjusted for inflation to $36 billion for 1990. The committee is convinced that basic research, properly supported and directed, holds the promise to decrease chronic illness and disability in old age, enriching the quality of life and maintaining vigor and independence in the later years.

A NATIONAL RESEARCH AGENDA ON AGING

Basic Biomedical Research

While the term *aging* can apply to changes occurring during any phase of life, *dysfunctional aging* (or *senescence*) refers to changes associated with accelerated morbidity and mortality rates during the latter phase of adult life. Biomedical studies on the disorders of senescence include not only specific disease states but also conditions not easily classified. Basic biomedical science can substantially improve the quality of life for older persons, and add to the list of social and medical advances in which the nation takes pride, advances that eliminated so many of the killing and crippling diseases of childhood and younger adults.

A number of puzzling questions about aging have attracted the interest of biomedical researchers. Why do humans differ so individually in aging changes of bone, brain, and heart? What preconditions for aging may be set earlier in life, perhaps even in utero? What is the role of gene-environmental interactions in the individual patterns of aging? The committee believes that understanding these processes will lead to interventions that may delay or even reverse disability.

Although many fundamental processes of aging are yet to be understood, progress already has been made in distinguishing between the process of natural aging and disease states associated with older age. Achieving further knowledge will require contributions from genetics, biochemistry, cell biology, neurobiology, and other disciplines. Equally important is the integration of aging research with other areas of investigation that historically have focused on specific organs and diseases without considering the surrounding manifold of aging changes. The committee emphasizes that many major advances in science have been unanticipated. Therefore, it is essential to maintain the traditional primacy of investigator-initiated studies—the source of many research breakthroughs.

Recent advances in the basic biomedical study of aging will contribute significantly to the foundation for future progress in the understanding of how individuals age. Among these achievements are insights into gene regulation and other basic cellular functions; the development of colonies of rats that have shown increased lifespan and delayed decline of function following food restriction; and the creation of strains of mice with targeted mutations (i.e., basic changes in the genetic blueprint for proteins) that will improve understanding of the processes of aging.

Major conceptual and methodological successes in molecular biology will lead to high-resolution gene mapping and to increased understanding of gene expression at the molecular level. These advances have the potential for identifying the location and cloning of genes responsible for various age-related disorders, including familial Alzheimer's disease. It will then be possible to study how such genes may be controlled to prevent or further delay the adverse effects of age-related disorders.

The elucidation of the biological mechanisms of aging and the clinical and lifestyle interventions derived from these discoveries will improve the quality of advanced age. Moreover, such research findings are eminently achievable.

Priority Research Recommendations

For the 1990s and the first decade of the twenty-first century, planning and implementation of a multiinstitutional basic biomedical research effort in two principal areas should be undertaken.

Abnormal cell proliferation Research in this area involves the study of proliferative homeostasis, a fundamental process responsible for the orderly replacement of cells lost because of either exposure to toxins or endogenous processes (Martin, 1979). Mechanisms underlying pervasive disturbances in proliferative homeostasis that accompany aging in humans and other mammals are pivotal in many diseases that cause senescence. Alterations in cell replacement are vital to regeneration and repair, but they contribute to the genesis of many forms of cancer, atherosclerosis, osteoarthritis, benign prostatic hyperplasia, and disturbed immune function. The committee proposes that this major new national initiative be comparable in emphasis to the number one priority of the NIA—Alzheimer's disease and the neurobiology of aging.

Brain aging Basic research in the neurosciences (including the peripheral and central nervous systems) and the current special research initiatives on Alzheimer's disease should be continued and expanded.

Additional Research Opportunities

The research priorities in basic biomedical science listed above merit primary consideration for implementation, but other promising areas of biogerontological investigation, also merit support. Some of these important research directions are as follows:

- the study of the repair and regeneration of postmitotic nondividing cells, including neurons and cardiac and skeletal muscle cells, as well as conditionally proliferating cells, such as hepatocytes;
- the study of the major integrative systems of physiology, including the effects of aging on the immune and endocrine-neuroendocrine systems, the effect of dietary restriction on lifespan, and the regulation of reproductive and developmental physiology;
- the study of the varying rates of aging among different cells, tissues, and organs for individual differences within species and for species differences that will provide information about genetic influences;
- the study of the biomarkers of aging, their relationship to overall functional state, and the role of functional capacity as a marker of senescence;
- the study of the general question of gene expression in aging (especially as applied to the abnormal proteins associated with Alzheimer's disease or the inactivation of enzymes with age), selective changes in gene activity, and the effects of environmental influences on aging (e.g., free radicals, radiation, and various toxins); and
- the use of comparative genetics and development of new model systems to study aging and the lifespan.

The committee emphasizes that additional resources to support research on aging should supplement, not supplant, existing support for current basic research, either in aging or in other areas of biological investigation.

Resource Recommendations

First in importance, increase the funding of NIA—and other NIH— approved investigator-originated research proposals on aging from the present level of one in four to one in two.

- Increase the availability of animal models (including primates and other long-lived animals that more closely resemble humans) and laboratory animal models with contrasting maximal lifespans.
- Improve access to cell lines and tissue samples, especially cryopreserved cells and tissues, and cell and tissue types from different human populations that crosscut age, sex, and stages of disease (e.g., different stages of diabetes mellitus or Alzheimer's disease).
- Expand existing human longitudinal studies, making available archival information from previous studies, including data on mortality rates and incidence of disease/disability from different human and animal populations.
- Add at least 10 Centers of Excellence in Geriatrics and Gerontologic Research and Training (Claude Pepper Centers) to the current three supported by the NIA (see funding section), in which multidisciplinary age-related studies can be carried out and in which high-technology tasks fundamental to a wide variety of research on aging can be supported.
- Increase support to train an additional 200 basic biomedical scientists per year in gerontology.

CLINICAL RESEARCH

The rapid growth of clinical research in geriatric medicine in the past decade has yielded new understanding of the physiology of aging, the mechanisms underlying many age-related disorders, the effectiveness of intervention to prevent or treat these disorders, and the characterization of frail elderly people, including those in long-term care.

The committee has reviewed this database to identify two major priority areas to serve as the basis for specific research questions and proposals.

These priority areas, described below, overlap in some instances with interests shared by basic biomedicine and other fields of research, including the interest of behavioral/social studies and health services delivery in the study of the effectiveness of intervention in improving the health and function of older persons. This overlap illustrates the interdisciplinary character of age-related research.

Priority Research Recommendations

Research into the causes, prevention, management, and rehabilitation of functional disability in older persons, including a focus on the geriatric *syndromes* Functional disability in older persons is the major obstacle to independence, and 23 percent of older persons are impaired (less for those under 75; about 50 percent for those 85 and over) in one or more of the self-care activities that make independent existence possible (National Center for Health Statistics, 1985). The clinical research community currently is well positioned to make significant short-term advances leading to improvement in dependency and frailty in persons of advanced old age. This group is increasing rapidly in size and needs help urgently, especially in health promotion and disease prevention, and in studies of specific geriatric syndromes.

Health promotion and disease prevention in old age Although many research areas have been identified, two are offered as examples that show special promise for discovery. First, research should be conducted on the interacting effects of age, lifestyle, and disease on disability among older persons. Intervention studies should determine the changing effect of such factors in the prevention and rehabilitation of disability. Second, studies should examine the effectiveness of various disease prevention strategies for elderly persons and the extent to which preventive interventions may forestall late-life disabling conditions that begin in earlier life. Obviously, this research is interdisciplinary, overlapping with social and behavioral issues.

Studies of geriatric syndromes Although many syndromes are age associated, the five listed below offer unusually promising opportunities to benefit from careful clinical exploration. These syndromes are among the most prevalent of all late-life maladies.

- Failure to thrive: This syndrome involves poor nutrition, including decreased appetite and weight loss (often with dehydration), inactivity, depression, impaired immunity, and low cholesterol. One promising line of research is the conduct of crosscohort and longitudinal studies to determine true prevalence, risk factors, and precipitating events of this syndrome. Another important area is the study of interventions to prevent or treat poor nutrition and other factors associated with failure to thrive, such as pressure sores, increased infection, and depression.
- Impaired postural stability, muscle strength, and mobility: This widespread problem includes falls, dizziness, syncope (fainting), fractures, muscle weakness, and lack of mobility. Each year over one million older persons sustain fractures, including 200,000 hip fractures, that are related to this syndrome (Cummings et al., 1985). Recommended research includes (1) the study of pathophysiologic mechanisms and other factors (e.g., poor nutrition, auditory and visual impairments, and specific neurological defects) underlying recurrent falls and (2) the study of interventions, including drugs, that improve or further impair some of these functions.
- Mismanagement of medications: Polypharmacy, defined as the taking of three or more medications regularly, occurs in one-third of those over 65. Research in this area should include (1) studies of the fundamental pharmacodynamics of medications in older individuals and (2) studies of medication use and effects that link discoveries in basic biomedicine (e.g., gene delivery systems, specific enzyme-blocking agents, and monoclonal antibodies) to clinical studies, behavioral investigation, and health services research.
- Urinary incontinence: Although the degree of urinary incontinence varies widely among individuals, this condition is present in approximately 30 percent of community-dwelling older persons (Diokno et al., 1986); 50 percent of those in nursing homes are incontinent (Ouslander et al., 1982). Urinary incontinence has major

consequences in suffering and in costs of care among the older population. Behavioral, neuroanatomic, and neurophysiologic studies are needed to determine the basis of urinary incontinence and the efficacy and risk of currently available treatment modalities.

- Delirium: About one-third of hospitalized elderly individuals develop acute confusion, markedly complicating their hospital course and dramatically increasing morbidity and health care costs (Lipowski, 1989). Very little is known about the predisposing factors, natural history, underlying mechanisms, and effective treatment of delirium. Studies of the interaction of age-dependent physiologic changes and *important diseases in old age* Research on the interaction between disease and aging is interdisciplinary, drawing on insights provided by research in molecular biology, physiology, nutrition, and behavioral and social science. This research may prove critical for issues of functional capacity, because the major diseases discussed below impose heavy burdens in old age and because better understanding of the linkage between age-related disease and disability may lead to improved independence in older people.
- Cardiovascular disorders: Cardiovascular disease is the major cause of death in older persons and is a leading cause of chronic illness and disability. Research should explore genetic and other risk factors as well as the molecular basis of atherosclerosis; the development of standard and recombinant-based drug technologies along with gene therapy approaches and nutritional interventions to treat or prevent hypertension and atherosclerosis; and evaluation over time of therapy for cardiovascular disease to assess cost benefits and effect on quality of life.
- Dementia and affective disorders: Dementia after age 75 affects 40 percent of the population (Hagnell et al., 1983) and is the major reason for disability among the very old. Great is the need for studies linking basic and clinical approaches to an understanding of the pathophysiology, cause, prevention, and treatment of Alzheimer's disease

and other forms of dementia. These studies should include assessment of cellular and molecular markers as well as new approaches to imaging and neurotransmitter mapping of the nervous system. Affective disorders are underreported in old age, although they account for substantial disability (Finlayson and Martin, 1982). Multidisciplinary studies to investigate the pharmacology, neuroendocrinology, and behavioral and social aspects of affective disorders are recommended.

- Musculoskeletal disorders: Musculoskeletal disorders are second only to cardiovascular disease both as a cause of disability and as a cause of high health care costs among older persons. The application of techniques of powerful reverse genetics such as recombinant fragment length polymorphism (through which genes that contribute to serious genetic disorders such as cystic fibrosis have been found) will help to define the genetic basis of osteoarthritis and osteoporosis in many patients with these common disorders. This approach should receive a high priority as gene replacement techniques are developed as potential cures of such disorders. Other recommended research includes study of nutritional factors and nutritional therapy, and the investigation of the effectiveness of traditional and new rehabilitative interventions on musculoskeletal disorders.
- Infectious diseases and immunosenescence: Although infection is felt to play an important role in disability in older patients, further study is needed on the contribution of infection to disability and on the effect of prevention and treatment of infection in the older population. Further examples of research include topics such as whether vaccines can bypass the need for immune T cells and whether vaccines made with live attenuated organisms, vaccines with antiidiotypic antibodies, or vaccines with antigens attached to a rigid molecular backbone are effective in elderly persons. Efforts should utilize such new technologies as genetically engineered vaccines and attenuated viruses in studies in this important area.

- Neoplasia (cancer): Death from cancer is increasing among older persons; indeed, cancer is the second leading cause of lost years of life in the geriatric population. Although accumulation of mutations with age plays a major role in the emergence of neoplasia, more knowledge is needed to explain why age is the most important predictor of cancer in this group; nor is there sufficient information about the value of prevention and treatment of cancer in older patients. Reverse genetics (as described above), recombinant DNA products, and delivery systems using gene transfer approaches (mechanisms for inserting DNA into the body) should be especially valuable in this area. Finally, research into nutritional and other risk factors in the development of cancer should be expanded.
- Disorders of metabolism and homeostasis: Disorders of metabolism and homeostasis may precede and almost always influence the emergence and severity of the conditions described previously. The committee recommends studies on the pathophysiology and molecular genetics of conditions such as diabetes mellitus, altered muscle metabolism, altered bone and mineral metabolism, and altered lipid metabolism in old age.

Additional Research Opportunities

These Include

- investigation of the molecular biology, pathophysiology, prevention, and treatment of common sensory deficiencies of old age, especially impairment of hearing and vision;
- examination of the factors associated with oral and dental problems of old age, including periodontal disease, disorders of salivation, and oral- facial pain;
- study of common skin conditions of elderly persons, such as dry skin, itching, and predisposition to cancers of the skin; and
- study of the interaction between nutrition and normal aging and between nutrition and age-associated diseases.

Resource Recommendations

The support required for an effective research program in the priority research areas involves added funds, phased in over 5 years, most importantly an increase in the funding rate of NIH-approved investigator-initiated research grants on aging from one in four to one in two; provision of education and support for an additional 200 to 300 fellows and postdoctoral students and 100 to 150 junior faculty per year, coupled with necessary continued support for midlevel investigators new to the fields of geriatrics and gerontology; expansion of current infrastructure supports; and the addition of at least 10 new Claude Pepper Centers, as described more fully in the section on funding.

BEHAVIORAL AND SOCIAL RESEARCH

Fundamental advances in the study of aging have shown that psychological and behavioral variables, along with factors in the social and physical environment, can alter the course of aging dramatically. Thus, the process of aging is highly responsive to social and behavioral interventions. Studies employing such interventions as training in cognitive skills or increasing the sense of competence and self-efficiency in some older individuals who are deficient in these qualities have had dramatic effects on function and health.

Research in the social and behavioral sciences has emphasized that aging is an interactive and lifelong process. This view also has provoked interest in the associations among biological differences, behavior and psychosocial processes, and alterations in physiological systems in health and disease.

The potential of the aging process to respond to modification provides a unique opportunity for future research. Individuals arrive at the end of life by quite different socially determined, as well as biologically determined, routes.

Differentiation in older adults provides some of the best evidence for the modifiability of aging processes and the experience of aging. *Differentiation* refers to variability in aging, both within and across societies. Studies have found that behavioral capacities are multidirectional over the adult life course: some remain stable, and others decline or improve. Because individuals adjust to biological decline in aging, behavioral decline often is minimal.

Aspects of differentiation in elderly persons also involve variations in individual and cohort lifestyles, risks of illness and dependency, and functional capacity. Cultural and psychological processes influence the aging process in varied and consequential ways. Most studies documenting this influence have focused on reducing age-related decrements in health. Research is needed to specify the broader areas and limits of the modifiability of aging, emphasizing the enhancement of existing skills and the potential for learning new skills in later life.

Another major perspective in social and behavioral research focuses on the interaction between person and environment. Specific sociocultural contexts influence the aging process differently in specific individuals. Study of these contexts should include how aging differs across societies and across cultural, racial, and ethnic groups; how aging is influenced by differences in work place, treatment environment, living arrangements, and other material and social supports; and how individual capacity and social milieu affect behavior and health outcome.

Priority Research Recommendations

Three research priorities emerge from these themes.

Investigation of the basic social and psychological processes of aging, including specific mechanisms underlying the interrelationships among social, psychological, and behavioral variables and between these *variables and biological aging functions, should be undertaken* Examples include investigation of the relationship between memory function and central nervous system structures such as between the hippocampus and cerebellum (Berger et al., 1986), and research on stress-related environmental, psychological, and hormonal factors that lead to anatomic changes in the brain (see cerebral atrophy following extreme stress: Jensen et al., 1982; Finch, 1987).

Special attention should be paid to at-risk populations, such as women, the poor, and minorities, since study of these populations may lead to new understanding of the dynamic interaction between biological and social variables. The relationship between brain and behavior represents an exciting new research frontier. Basic research is needed to provide a clearer understanding of individual differences in sensory, cognitive, and behavioral aging, particularly as these differences reflect relationships between brain and behavior, environment and behavior, and society and behavior. Clinical research should apply what is known about the modifiability of risk factors, skills, learning, and memory. Despite recognition that the major health problems of older people are chronic, little attention has been paid to behavioral and social interventions to reduce disability and provide new strategies for management. Research should address rehabilitative strategies, evaluate the social and emotional barriers to rehabilitation, and explore issues related to compliance.

Research that addresses issues of change in population, dynamics, and particularly the question of postponed morbidity is a further high priority Descriptive demography, epidemiology, and population estimates are especially important to forecast the number of older people who will be independent or not. New measures are needed to evaluate the sequence of disease and/or disabling impairment and death for individuals and social groups—compared by cohort over time. Research on the postponement of disability and dependence through changes in social context, and through behavioral intervention, must accompany studies of psychosocial predictors of health, longevity, and functional disability in the aged.

Psychological and behavioral variables may not only contribute to biological aging, but may also serve to mirror physiological dysfunction. In developed societies most older people are vigorous and independent, but certain groups of older adults (e.g., women and minorities) are at higher risk for poverty, social isolation, underemployment, inadequate education, illness, and inaccessibility to health care. Factors that postpone morbidity may function differently in these groups, making them especially important to study.

Research should be undertaken to study the manner in which societal *structures and societal changes affect aging* Just as individuals change, so too do social structures. Societal old age today shows neither what old age was like in the past nor what it will be like in the future. It is crucial to determine how stability and change in social structures such as the family and work place affect the performance, productivity, health, and quality of life of older adults.

Retirement from work or major changes in family structure or in family roles such as the increased entry of women into the workforce provide an opportunity to study how social change affects the experience of aging and health and the well-being of older persons. Research should examine the implication of these changes for social support and caregiving.

Research is needed as well to specify what changes in social structure could improve productivity in older people. Changes in mandatory retirement laws urgently call for more understanding of productivity and psychological processes in older workers. Finally, a life-course perspective should be adopted that views younger adults as the older adults of the future. If children and younger adults who are underemployed, undereducated, and underinsured can be identified and helped, this may improve the well-being of the older population of the future.

Studies derived from the priority research areas identified above may involve large-scale, longitudinal investigations. At the same time, consideration should be given to exploring the use of small-sample studies to enhance our knowledge on key aspects of the aging process.

Additional Research Opportunities

Opportunities for social research include the following:

- study of the characteristics of employment, work places, and older individuals associated with continued productive activity in lifelong jobs or in new careers;
- study of how older people are affected (expected and actual performance) by changing technology, and how firms use training and job respecification to make technological changes as age neutral as possible; and
- development of a quality of life index to identify contributing factors among subpopulations for whom the index is low (e.g., the very old and minorities).

Opportunities for behavioral research include the following:

- study of the psychological concomitants of illness and how these affect self-care and response to formal care;
- study of the comparative effectiveness of different modalities in the treatment of chronic mental disorders; and
- examination of the effect of behavioral and social intervention on the outcome of long-term illness.

Resource Recommendations

Despite increases in the federal commitment to health research, the social and behavioral sciences have lost ground relative to other areas of investigation. Support for behavioral and social research on aging in 1989 was estimated at $80– $100 million from the federal government and $10–$15 million from nonfederal sources such as foundations (Behavioral and Social Research Program, NIA). Estimates of resources for the research agenda primarily involve investigator- originated studies and are to be phased in over a 5-year period.

Behavioral and social research will require substantial added funds during the first 5 years of the new research agenda, especially to raise funding for approved NIH research proposals on aging from one in four to one in two. Because much of the support for behavioral and social studies in aging comes from agencies outside the NIH (e.g., the Health Care Financing Administration and the Alcohol, Drug Abuse, and Mental Health Administration), support for research on aging by non-NIH agencies should be increased by at least 50 percent.

Other support includes training for 200 more behavioral and social scientists per year, one-time costs for construction of additional centers for multidisciplinary research (see the recommendations for funding section), and a share in funds for infrastructure utilized by all areas of research on aging.

Health Services Delivery Research

The growth of the older population, particularly those 85 and over, the chaotic state of today's health care system, rising health care costs, and mounting public concern about value received for the health care dollar lend urgency to the need for a new impetus in health services delivery research. Research in health services delivery expands knowledge about the organization, financing, and delivery of health care services—the path that connects older persons to the benefits of basic biomedical science, clinical studies, and behavioral and social research.

Major problems and issues of health services delivery for older persons

include:

- continuity of care and long-term care, especially the uneven distribution of services and financing for older persons and the ill- defined relationship between acute and chronic care;
- drug therapy, especially inappropriate prescribing of medications; poor compliance; and unappreciated alterations in pharmacokinetics and pharmacodynamics in older patients;
- mental health services, particularly underutilization of mental health services by older people, and the needs of mentally ill elderly persons in nursing home settings; and
- disability/disease prevention and health promotion services, most importantly the lack of information about the distribution of specific risk factors and the effectiveness of risk reduction programs in older persons.

Priority Research Recommendations

Research in the area of long-term care and continuity of care Research in long-term care and continuity of care should examine cost effectiveness of care; factors determining the need for and utilization of care; the organization and delivery of care (including the special role of the family); the adaptation of technology to foster independence of older persons; caregiving institutions (hospitals, nursing facilities, and other sites); factors determining the locus of patient care; and factors influencing the recruitment, training, retention, and quality of providers of formal and informal long-term care.

The costs and financing of long-term care Research is recommended on existing barriers to care of older persons, the requisite financial elements of insurance programs to support long-term care, mental health services, rehabilitation, and disease prevention and health promotion.

Studies of drug use Investigations should examine factors that influence the development of drugs and the regulatory approval of drugs; the efficacy and efficiency of prescribing, dispensing, and taking of pharmaceuticals; inappropriate drug therapy; interventions to maximize drug treatment; and examination of drug-related phenomena affecting social and physical functioning.

Mental health services research This research should be expanded to identify the determinants of the need for and use of mental health services, the role of treatment as a factor in improving social and physical functioning, and the effect of earlier psychiatric intervention on reducing later-life disability and health care costs.

Research in the fields of disability/disease prevention and health *promotion* There is a major need to develop new models for the delivery of appropriate behavior change programs for elderly populations.

Additional Research Opportunities

These include:

- study of alternative methods of delivering long-term care, including models that focus on a particular subgroup (e.g., posthospital management) and that link housing arrangements with community long-term facilities;
- examination of the effect of employer-initiated cost-sharing on the use of preventive and treatment services;
- study of the effectiveness of case-management compared with other methods of delivering and organizing health care;
- research on the provision of mental health services to elderly institutionalized persons in the context of need, utility, and outcome;
- investigation to compare the relative contribution of accepted risk factors in the development of disease among

young and aged populations; and

- research into the effectiveness of new technologies, including information technologies (computers) and robotics, in the delivery of health care.

Resource Recommendations

To implement the health services delivery research program the committee recommends additional funds to (1) provide support for research grants on aging to increase the current level of support for approved grants from one in four to one in two, (2) train an additional 140 investigators per year in the health services delivery field, and (3) increase databases and meet infrastructure needs. Health services delivery research will also participate in the 10 new multidisciplinary research and training centers (see the funding section), and in shared infrastructure funding. Because significant funding for health services delivery studies comes from outside NIH, additional funds are requested from agencies traditionally supporting this research (e.g., the Health Care Financing Administration and the Agency for Health Care Policy and Research).

RESEARCH IN BIOMEDICAL ETHICS

Ethical issues accompany all aspects of health care of older persons and are at the heart of research on this group. Areas of major importance in ethics and in the study of aging include the following: dilemmas regarding life- sustaining treatment, selection of therapeutic interventions based on age, distribution of health care resources, and the need to include older subjects in research. Prolonging life in some patients may not (or may) be desired by them or be appropriate; guidelines to establish responsibility for these decisions often are problematic for elderly persons.

Competing social needs, growing costs of care, and the enlarging population of sick and disabled older persons pose problems of resource allocation that require agreement on how to make such determinations fairly. Finally, there is a need to determine the most ethical way to do research on older persons who are institutionalized, frail, or cognitively impaired.

Priority Research Recommendations

Funds should be provided to conduct analytical and empirical research on biomedical ethical issues in three priority areas:

Dilemmas regarding life-sustaining treatment

Allocations of health care resources

Participation in clinical research by frail elderly persons

Additional Research Opportunities

These opportunities include:

- study of clinician-patient interaction regarding life-sustaining technologies;
- research into decision making for incompetent patients who lack advance directives;
- study of the role of institutional ethics committees;
- the identification of medical futility;
- study of resolution of disagreements between caregivers and patients or their families;
- defining appropriate care standards for elderly persons;
- investigation of the use of age as a criterion for allocation of health services;
- study of ethics in day-to-day interaction between caregivers and older patients (e.g., autonomy of older patients); and
- research into the trade-off between quality of care and quality of life for older patients.

Resource Recommendations

Although the NIH National Center for Human Genome Research recently set aside 3 percent of its funds for the study of ethical issues, which could include ethical issues in the care of older persons, the committee knows of no other federal support for research on ethics and geriatric patients, and it strongly encourages funding in this area, following the model of the genome project in association with specific biomedical, clinical, social and behavioral, and health services delivery research projects.

CROSSCUTTING ISSUES

Crosscutting issues bridge the disciplines and call for an interdisciplinary approach to their study. Such issues include gender, ethnic origin, cultural background, ethics, race, and the interdisciplinary approach to training and scientific investigation.

BASIC BIOMEDICAL RESEARCH

Major crosscutting issues here involve the study of the effect of race, gender, ethnic background, and other factors on the trajectory of aging, ranging from the cellular level to the intact organism. This study involves coordinated efforts with other disciplines to provide an integrated approach to multifactorial phenomena. The overarching issue of ethics applies to basic biomedical investigation in raising questions about the care and disposition of experimental animals and the application of new discoveries, such as gene transplants. Finally, interdisciplinary education in the area of gerontology/geriatries should be supported.

CLINICAL RESEARCH

Clinical research on aging engages many disciplines. Insights from molecular and cell biology provide a basic biomedical foundation for much clinical research. Behavioral and social research along with health care delivery studies can facilitate clinical investigation of the response of older people to illness and of the effect of interventions to reduce disability and morbidity occurring late in life.

Of equal importance are crosscutting issues that involve gender racial, and ethnic differences in response to illness, including how older persons experience illness, or in the metabolic disposition of drugs.

It would be interesting to understand, for example, the marked difference in morbidity of blacks compared to whites with equal degrees of hypertension or of women compared to men insofar as hypertension-related morbidity is concerned. Another issue involves ethics in clinical research—a subject discussed in the previous section of this summary.

BEHAVIORAL AND SOCIAL RESEARCH

Many crosscutting issues arise in behavioral and social research, including the effects of gender, race, and ethnicity on longevity, functional status, and morbidity in old age; social role expectations and appropriateness of intervention; and population dynamics to predict longevity and morbidity. In addition, neurobiological study of cognitive defects will be enriched by combination with approaches to intellectual function and its modifiability through behavioral and social intervention. Study in these areas requires interdisciplinary research that cuts across all of the areas of study described in this report, from basic biomedical to health services delivery.

HEALTH SERVICES DELIVERY RESEARCH

Significant differences in the utilization of health services by men and women extend beyond the distribution of chronic illnesses and morbidity in these groups. Gender differences are seen in participation in programs of prevention and in economic/social status of older persons. These variances should be investigated further. Other topics include the assessment of need for health services based on different patterns of illness and disability between men and women, gender influence on response to different modes of health care delivery, and sex differences in utility of common preventive and risk reduction interventions.

RECOMMENDATIONS FOR FUNDING THE RESEARCH AGENDA ON AGING

The committee believes that a major investment in research on aging is needed urgently and therefore recommends additional funds of about $312 million per year, not adjusted for inflation, phased in over a 5-year period. These funds, added to current research supports of approximately $601 million, total $913 million annually, a figure approaching the $1 billion recommended by the Pepper Commission (1990). In addition, a nonrecurring expenditure on construction of about $110 million will be required. Programs of research on aging and support for centers should continue to be funded on the basis of merit and the accepted high standards of competition for research funds. To estimate added funds for the study of aging the committee used available data (e.g., costs for centers, research projects, and training); where substantiating evidence was limited, the committee's decisions were based on judgment.

The federal government bears most of the cost of health care for older persons and is the only institution with the resources and organization to provide the wide-ranging research support requested in this report.

Foundations and industry will add to support for research on aging, but it is mainly to the federal government, especially for its support of NIH and other federal department and agency sponsored research on aging, that the committee's recommendations are directed.

It should be noted that in estimating funds for age-related research the committee did not consider the effects of a large increase in spending on research on aging on other U.S. health and human welfare-related research, and the committee realizes that resources are not unlimited. In addition, the committee does not comment on the desirability of increasing expenditures on research on aging versus expenditures on research on children, cancer, AIDS, contraception, the brain, or other major areas of investigation involving health care of younger adults.

Although this report is not intended to set priorities in the general area of health research, or to set priorities for all health research, the recommendations for additional funding in this section follow from the committee's conviction that increased fundamental research on aging holds the most promise to improve the lives of the ever-increasing numbers of older Americans.

Table below summarizes the National Research Agenda on Aging. The additional resources required to develop this agenda include research project grants, centers for the study of aging, funds for infrastructure, construction for centers and infrastructure, and training of new investigators.

Funds required to obtain these resources are summarized in Table below. Additional funds to support the research agenda have not been adjusted for inflation and are to be phased in over a 5-year period.

TABLE: National Research Agenda on Aging: Summary of Research Priorities

Research Area	Research Priorities
Basic biomedical research	Abnormal cell proliferation
	The aging brain
Clinical research	Functional impairment and disability
	Interaction of age-dependent physiological changes and disease
Behavioral and social research	Interaction of social, psychological, and biological factors in aging
	Changes in population dynamics; the postponement of morbidity
	Changes in societal structures and aging
Health services delivery research	Long-term care and continuity of care
	Costs and financing of long-term care
	Medications and older persons
	Mental health services
	Disability and disease prevention and health promotion
Biomedical ethics	Dilemmas involving life-sustaining treatment
	Allocation of health care resources
	Participation of older persons in research

SOURCE: Institute of Medicine, 1991.

DERIVATION OF RECOMMENDATIONS FOR FUNDING

Funds for Research Projects

Major achievements in the health sciences are rarely predictable, but have usually come from the laboratories of gifted investigators exploring basic research questions. These advances often require the support of research originating from individual researchers or from small groups of investigators. Therefore, the preeminent expansion of support for the research agenda on aging must be a major increase in funding for approved age-related research project grants (RPGs), especially the traditional investigator-initiated proposals (R01s) by the NIA and other branches of the NIH.

TABLE: Summary of Additional Funds (rounded off in millions of dollars) Needed for the Research Agenda

Resource Support	Funds Needed
Increased NIH funding of research project grants	$ 92
Funding for cooperative clinical trials	25
Non-NIH research support for behavioral/social and health services delivery research	80
Increased support for training of scientists	50
Infrastructure	50
Centers of Excellence	15
Total	$312
Construction	$110

NOTE: Funds are for yearly funding of resources, except for construction funds. Construction funds are a one-time cost. All estimates are in current dollars.

In fiscal year 1990 the average funding of approved RPGs at the NIH was 25 percent (4,917/18,956); at the NIA funding of approved RPGs was 24 percent (192/799)(Office of Information Systems Branch, Division of Research Grants, National Institutes of Health). The percentage of funding for approved RPGs is slightly lowered by the presence of reapplications. For example, in fiscal year 1990 at the NIA 30 percent of the approved RPG applications (241/799) were resubmissions.

The NIA funding rate for approved reapplications was 23 percent (55/241), and for first-time applications was 25 percent (137/558); the lower rate of funding of reapplications slightly diminished the overall funding rate of approved RPGs to 24 percent. A similar small effect of reapplications on the percentage of funding of approved RPGs was seen for the total number of RPGs funded by the NIH.

To improve support for this most important source of discoveries in aging, the committee recommends that funding of approved RPGs in aging by the NIA and other divisions of the NIH should be increased from the current level of one in four to one in two. In addition, although the U.S. House of Representatives has recently advised that the average NIH funding cycle be limited to 4 years, because of the long duration of chronic illness and disability characteristic of aging, the committee urges that more review cycles for research on aging be extended to 5 to 7 years. To increase the funding of approved RPGs on aging, the committee recommends an additional $92 million per year for this purpose, to be phased in over 5 years.

The committee suggests that a significant part of this increase should be made available as soon as possible in order to take advantage of high- quality projects ready for implementation and also to provide early encouragement to scientists interested in entering the field of studies on aging. The increase in funding applies to RPGs at the NIA and to RPGs on aging approved by other branches of the NIH.

In fiscal year 1990 NIA support for 192 RPGs (at $190,000 per RPG) was $36.5 million, or about 17 percent of the total NIA budget of $239 million (Budget Office, NIA). The percentage of funds devoted to RPGs at the NIA was higher than the overall 12 percent ($945 million/$7.6 billion) of the total NIH budget funding of all RPGs for fiscal year 1990 (Division of Research Grants; Reports, Analysis, and Presentation Section, NIH).

The committee was unable to obtain information about the number of non- NIA-funded RPGs on aging, but estimated a minimum number of about 294 RPGs. This estimate was based on an average cost of $190,000 per RPG per year and assignment to age-related RPGs of 30 percent, or $55.8 million, of the $186 million committed to research on aging by non-NIA institutes in fiscal year 1990.

This is a conservative estimate: the actual funding and number of these non-NIA RPGs may have been higher because the non-NIA institutes committed fewer dollars to training and other non-research project supports of studies on aging than did the NIA. Given the estimate of non- NIA support for RPGs on aging, the figure of about $92 million in additional funds can derived from the following calculations:

Funding of approved RPGs on aging: funding level—one in four

- NIA $36.5 million (192 RPGs)
- NIH $55.8 million (294 RPGs)
- Total $92.3 million (486 RPGs)

Funding of approved RPGs on aging: funding level—one in two

- NIA $36.5 million × 2 = $73 million (384 RPGs)
  $36.5 million additional funding (192 added RPGs)
- NIH $55.8 million × 2 = $111.6 million (588 RPGs)
  $55.8 million additional funding (294 added RPGs)
- Total $92.3 million additional funding (486 added RPGs)

These figures are estimates and may require revision as the new agenda on research is implemented. Because a significant percentage of RPG proposals represents resubmissions that were previously approved but not funded, the new funding rate for approved grants may have variable effects on the eventual number of approved grants that are reviewed for funding. Increasing the funding of approved research grants on aging should reduce the number of resubmissions that are reviewed for funding, but this trend will be offset if more investigators are encouraged by the new funding rate to resubmit grant proposals that were approved but not funded in the past. Therefore, the committee recommends that the target for funding of approved RPGs be set at one in two, subject to review and modification as the new program to expand research on aging is implemented.

Funds for age-related research sponsored by the NIA and other institutes most often support biomedical research proposals, although some funds also support research on aging in health services delivery and in social and behavioral studies. As Chapters 4 and 5 emphasize, these areas have been chronically underfunded for many years. For example, an estimated fewer than 20 percent of approved research proposals on aging in behavioral and social research are funded each year (Behavioral and Social Program, NIA). The committee recommends that most of the funds for the behavioral and social research and health services delivery research should come from agencies that have traditionally sustained this work (e.g., Health Care Financing Administration; Alcohol, Drug Abuse, and Mental Health Administration; and the Agency for Health Care Policy and Research), and that the new funds should total at least $80 million per year in current dollars.

The 1991 NIA appropriation funds 381 RPGs (Budget Office, NIA). This new support is in concert with the committee's recommendation for increased funding of RPGs on aging, and should be extended to RPGs on aging throughout the NIH. There is information that other federal institutions may also increase their support for age-related research in the coming fiscal year. Funding for an additional five cooperative studies on prevention, demography, epidemiology and treatment of age-related disorders is recommended by the committee. These studies would implement research priorities identified earlier, and should be phased in over a 5-year period. Cooperative studies require about $5 million per study for support in current dollars (Director's Office, NIA), totaling $25 million dollars at full implementation

Funds for Training in Age-Related Studies

A recent IOM study recommended that the number of graduates in medical academic geriatric programs be increased from its current level of 100 per year to 200 to 250 per year to meet the estimated 2,100 biomedical faculty members needed in this field by the year 2000 (IOM, 1987; Rowe et al., 1987).

Additional funds should be phased in over a 5-year period to implement this recommendation. Assuming that training programs for clinical investigators last 2 to 3 years, this would involve support for an additional 200 to 300 fellows in training at all levels. In addition, funds should be provided for 100 to 150 beginning clinical investigators in research on aging.

Because training programs in age-related research are currently undersubscribed, it may be best to phase in major support for training programs *after* it has been shown that the increased investment in research has attracted more students to the field. The committee believes that the current cadre of faculty members with interests in age-related research is adequate to provide initial training for increased numbers of students in this area.

The NIA estimated in 1987 that the number of trainees in age-related studies, including clinical and basic biomedical research on aging, was far short of the number needed to meet present needs (NIA, 1987). The committee recommends that steps to repair this deficiency include support for 200 additional trainees per year in basic biomedical science.

Training in behavioral and social studies and in health services delivery research has long been poorly supported (Chapters 4 and 5). In a 1987 report to Congress, the NIA predicted that more than 1,500 nonbiomedical faculty, including behavioral and social scientists, would be needed by 1990 for teaching and research needs in the field of aging, and that more than 3,500 such professionals would be needed by the year 2000 (NIA, 1987). The report noted that by 1987 only a small percentage of the 1990 training needs had been met. The committee's recommendations provide for an additional 200 doctoral trainees per year in behavioral and social studies in aging, and 140 additional trainees per year in health services delivery research.

Acknowledging that there are differences in stipends, benefits, and faculty support, depending on the level of training and the discipline involved, the committee judges that the added training across the areas of study will involve stipends, benefits, and faculty support at an average cost of $50,000 per trainee. The total number of additional trainees recommended is about 1,000, and the cost —not adjusted for inflation—for this training is $50 million per year, phased in over 5 years.

Funds for Centers

The committee recommends that funds should be phased in over the next 5 years for 10 Centers of Excellence in Research and Teaching in Geriatrics and Gerontology (Claude Pepper Centers) to be added to the current 3 NIA-sponsored Centers of Excellence.

Although the value of the centers has been questioned, and there is no clear understanding of the number of centers needed, arguments for interdisciplinary centers on aging have been given in a recent IOM report (IOM, 1987), and by the 1984 Department of Health and Human Services report on education and training in geriatrics and gerontology. According to these reports, during the early phases of the evolution of the study of aging, centers can (1) serve to mobilize and focus scarce resources, (2) provide a setting for growth of new investigators, and (3) enhance creative interaction among scientists in diverse areas of research on aging (IOM, 1987).

To merit support for the development of a Center of Excellence, "a program must meet three central goals: 1) to develop a structured, efficient training program that will attract students and produce faculty; 2) to conduct research to add to clinical knowledge, maintain an academic base, and provide role models for trainees; and 3) provide clinical care in a variety of settings. . . . "(IOM, 1987). If these goals are met and if the presence of a cadre of investigators and teachers (whose research efforts are already well supported) can be demonstrated, then a university or medical center may qualify for center support. The NIA currently supports three centers. Four additional Centers of Excellence have been approved, but not funded, and six other sites are under consideration for development. Review of applications received by the NIA indicates that sufficient faculty numbers now exist to provide the necessary professional staffing of these proposed centers (Office of the Director, NIA). The committee believes that the new centers should be interdisciplinary in character, offering opportunities for exchange of ideas and collaboration across the fields of study of basic biomedical science, clinical research, behavioral and social studies, health services delivery, and biomedical ethics.

Estimation of the cost of the new centers is based on outlays for the Claude Pepper Geriatric Centers. These institutions required about $1.2 million per year for support (Office of Geriatrics Branch, Biomedical Research and Clinical Medicine Program, NIA). Based on this information, the 10 additional centers should cost about $12 to $15 million per year to operate (not adjusted for inflation). Funding of the new centers goes largely toward administration costs, for infrastructure costs, and toward other supports, such as salaries for beginning investigators. Direct funds for research would not depend on center support, but would come rather from grants to individual investigators and to research programs operating within the centers.

Funds for Infrastructure

Factors that make it difficult to assign costs for infrastructure research support to the different areas of research include the use of animal colonies by several disciplines for research, overlap of funds for research and infrastructure, development of databases that all disciplines may draw upon, and uncertainty in classifying new linkages to existing sources of information.

Given these constraints, the committee offers the following estimates for infrastructure costs.

Animal colonies Based on the cost in the Pepper Centers of $375,000 per animal colony per year (Office of Geriatrics Branch, Biomedical Research and Clinical Medicine Program, NIA), not adjusted for inflation, the estimated yearly costs of animal colonies at the 10 proposed geriatric research centers are $3.75 million; adding 10 animal colonies at other university centers engaged in gerontological research brings the total cost to $7.5 million per year.

Laboratories There are no comprehensive data on the cost of laboratories for basic biomedical and clinical research. Apart from laboratories at centers, funds are necessary to support noncenter laboratory needs in age-related basic biomedical and clinical research in other institutions. For example, three sophisticated laboratories for cell biology research to serve as regional resources for basic biomedical and clinical scientists would cost $3 million per year to staff and maintain (G.M. Martin, University of Washington, personal communication, 1989).

Databases This important infrastructure will be shared by all disciplines and consists of ongoing population studies, improved linkage to existing sources of information (e.g., Medicare data tapes), library supports such as the National Library of Medicine, and computer technology.

Research in geriatrics and gerontology will require the expansion of existing population studies (e.g., National Health and Nutrition Examination Survey) and the institution of new longitudinal demographic and epidemiological studies. The cost of such studies covers a wide range, from $1.5 million to $2.0 million per year (Office of the Director, NIA) for the Health and

Retirement Study funded by the NIA, to $5 to $10 million per year (G.M. Martin, University of Washington, personal communication, 1989) for large cardiovascular prospective studies. The committee estimates that population studies would come to about $30 million per year. Another $10 million per year, shared among the disciplines, would be needed to add computer technology, to fund improved access to existing databases, and to increase the capacity of reference sources such as the National Library of Medicine.

All told, the estimated nonassigned infrastructure support totals $40 million per year. Total costs for infrastructure, not adjusted for inflation, are estimated at $50.5 million per year, phased in over 5 years.

Funds for Construction

Construction costs include those for 10 additional multidisciplinary centers and those for out-of-center animal facilities and laboratories. The cost of construction of the Mental Retardation Research Centers supported by the National Institute of Child Health and Human Development was, on average, $10 million per center; the federal government provided 75 percent of the funding, and local and state organizations made up the difference. Using these figures, with the federal government providing 75 percent of the cost of construction, the committee estimates that the total one-time cost for 10 new geriatric research centers will be $75 million, not adjusted for inflation, phased in over 5 years. Center costs include animal housing.

Additional animal housing will cost $300 to $400 per square foot (Office of Resources Development, NIA). If 10 animal facilities are built outside the centers, and if total floor space at each of these additional institutions comes to 1,000 square feet (assuming an average of 5 separate projects at each institution), total costs will be about $3.5 million. Costs to construct new laboratories, based on the committee's best estimate, are assessed at an additional $30 million, not adjusted for inflation.

The committee therefore estimates that the total one-time costs for construction, not adjusted for inflation, will be about $110 million; as described above, these costs are to be phased in over a 5-year period, with $33 million of this sum assigned to construction of infrastructure supports distributed among several areas of research.

Table 3 demonstrates the effect of low, intermediate, and full funding levels on implementation of the research agenda.

For example, funding at 20 percent to 40 percent of the recommended level, phased in over 5 years, would add only about 20 to 40 age-related research projects per year, or less than 10 percent to the total number of approved and funded research project grants on aging each year. Support for trainees over 5 years at this level would add only 40 to 80 trainees per year, or 4 percent to 8 percent per year of the additional number needed. To implement research on aging in the coming decade, funding levels for resources should approach or equal the recommended total of $312 million per year, plus construction monies.

Recommendations for funding outlined in this section were based, as far as possible, on known costs for construction, infrastructure, grant support, research centers, and training. Determination of funding needs will require further examination and review prior to implementation of the new research agenda. The estimates given here represent a first step in this process.

IMPLICATIONS FOR FUNDING AGENCIES

Support for the research agenda can come from many sources. The committee suggests the following:

- **Federal agency support: Support by the federal government has been detailed in the section describing funding requirements for research. As noted, large-scale projects require federal support. Additionally, the federal government should provide research funds for research in biomedical ethics. Early in this effort, the NIA and/or other agencies should convene a national conference to determine the need for support, levels of funding, and ways of implementing the research agenda on biomedical ethics.**
- Private and corporate foundations: The increased flexibility of foundations makes them better able to support

innovative research programs—both basic and applied—with relatively short-term goals. It would also be appropriate for foundations to join in supporting fellowships to train scientists for age-related research and to participate in the funding of geriatric research centers. Foundations can further support research in geriatrics and gerontology by funding national conferences to plan for research and dissemination of research findings or by funding training programs such as the Brookdale Foundation support of the NIA Summer Training Institute for young researchers.

TABLE: *Scenarios of Partial (Low, Medium, and High) Support for Aging Research*

	Scenarios (in millions of dollars)		
Research Support	**Low**	**Medium**	**High**
Total yearly funds	$62–115	$156–215	$250–312
(% of recommended)	(20–37%)	(50–69%)	(80–100%)
NIH support for RPGs (additional	$18–37	$46–64	$74–92
number per year.)	(97–194)	(243–340)	(389–486)
Cooperative clinical trials (additional	$5	$10–15	$20–25
number per year)	(1)	(2–3)	(4–5)
Non-NIH support for behavioral/social and health services delivery research	$16–32	$40–56	$64–80
Training positions (additional number funded	$10–20	$25–35	$40–50
per year)	(200–400)	(500–700)	(800–1000)
Centers of Excellence	$3–6	$7–10	$12–15
(additional number funded)	(2–4)	(5–7)	(8–10)
Infrastructure	$10–20	$25–35	$40–50
Construction	$22–44	$55–80	$88–110

NOTE: Totals may not equal sums of columns because of rounding off.

Funds are for yearly funding of resources, except for construction funds. Construction funds are a one-time cost. All estimates are in current dollars.

- Industry support: Resources to support new technologies, evaluate new drug therapy, and develop educational programs could be provided by industry sources.
- Combined support: Government, foundations, and industry could combine resources for nationwide policy conferences on age-related research, support for fellowships, development of geriatric research centers, and creation of functions to facilitate communication about studies in aging among different institutions. A recent IOM workshop recommended that planning be undertaken to explore and implement collaboration between government and industry in biomedical research and education (IOM, 1989).

CONCLUDING COMMENTS

The committee notes that the community of health science researchers is a great national asset—unmatched in number and productivity. Moreover, the cadre of scientists working in aging has increased rapidly throughout the country over the last 10 years and provides a major resource for new advances in basic and applied research. Prospects are bright for the application of new knowledge in the service of improved social, psychological, and physical health for older persons, for those who will reach old age in the near future, and especially for those who are now young. The numerous areas of basic and applied research in aging presently led by U.S. scientists also are a foundation for major innovation and application in the private sector. There is great potential for new drugs and consumer goods adapted to the special needs of older persons; the world market for them is growing rapidly. Moreover, the principles of genetic engineering already exist not only for organ repair and replacement but also for control of the adverse effects of inherited genes. Thus, the goals of this 20-year plan for research point to a unique U.S. role in meeting the needs of the older population throughout the world.

2

Population Trends in Recent Decades

Over the past two decades the American system of health care and the biomedical, social, and behavioral sciences have begun to come to grips with the overwhelming consequences of the rapid and dramatic growth of the elderly population. Long neglected, the topic of aging has attracted substantial attention as this country seeks to build the infrastructure of a national research and academic capacity in aging. For instance, the National Institute on Aging (NIA) and scores of university and medical school academic programs in gerontology and geriatrics have been established, and recently developed training programs are generating a cadre of sophisticated gerontologists and geriatricians. In addition to building the foundation for what is becoming one of the nation's major new scientific efforts, recent financial investments in aging have already yielded important research findings across a spectrum of inquiry, including the basic mechanisms of aging, the physiology of normal aging, the social and behavioral aspects of aging, health services research for older persons, and the etiology and management of age-related disease.

A bibliometric search, conducted in conjunction with the project that forms the basis of this report, shows that between 1983 and 1987, taking into account overall growth in research, there was a 70 percent relative increase in scholarly publications in aging. Although growth in some of these areas, particularly in age-related disease (e.g., Alzheimer's disease), has clearly outstripped growth in other areas, all domains have seen substantial growth from their meager beginnings over four decades ago.

Notwithstanding this progress, the medical and scientific community's knowledge, research, and clinical capacity lag far behind the human need, and society currently is paying a heavy price for decades of neglect of aging and the special needs of older persons. The burgeoning elderly population will increasingly burden the nation's fiscally troubled health care system. Despite our recognition that investments in research ultimately yield substantial financial, clinical, and social benefits, the nation's commitment to aging research is minuscule in comparison to the costs of health care for older persons. Even in the case of Alzheimer's disease—one of the best-supported areas of age-related research—expenditures for scientific investigation equal less than one-half of one percent (0.1 percent) of the costs of care for the victims of this disease (Report of the Advisory Panel on Alzheimer's Disease 1988–1989, U.S. Department of Health and Human Services, DHHS Pub. No. (ADM)89-1644).

Given the heavy toll and long duration of chronic illness and disability, the small national commitment to aging research is a particularly risky and potentially wasteful strategy, especially in light of the enormous promise of properly supported research, not only to improve the status of older adults, but also to reduce substantially the costs of their care. Effective strategies against many disorders common in old age, such as dementia, urinary incontinence, osteoporosis, and falls, could yield huge financial savings and, at the same time, enhance well-being. Likewise, research is needed on intervention strategies to ameliorate the deleterious conditions of economic and social dependency, the growing caregiving burden, and inadequate coping capacities. Such research could reduce negative health effects and their attendant costs while improving the aging experience for millions of Americans.

A factor limiting the development of gerontologic and geriatric research is the critical shortage of workers in this area. Current funding mechanisms often do not provide adequate research or training support for young investigators interested in studying aspects of aging. The United States must develop effective mechanisms to attract established scientists from other fields into aging or age-related research, and it must construct attractive career ladders for young investigators from a variety of fields who enter the study of aging as a career.

A growing sense of urgency mandates that the time is right to narrow the gap between the needs of an aging society and the scientific knowledge base. Building on the existing academic sub strata, the medical and scientific community can leverage both new knowledge and tools from other fields to develop aging research into a mature component of the national scientific portfolio. For example, dramatic recent developments in several fields—from molecular biology to the social sciences—await application to age-related issues and hold the potential of rapidly advancing the nascent fields of gerontology and geriatrics. There is an increasing awareness that the field of aging may be entering a phase of major breakthroughs that have the potential not only of increasing our understanding of age-related diseases and enhancing functions in the elderly but also, quite possibly, of beginning to solve the mystery of the human lifespan. This opportunity is constrained, however, by increasing fiscal pressures.

At this critical juncture the nation's needs demand a comprehensive plan for research on aging. This document represents the result of the Institute of Medicine's (IOM's) effort to develop a plan in recognition of this need at the request of a number of federal agencies and foundations.

The IOM convened 87 experts from various fields relevant to aging and age-related research in a two-year study to identify the needs and opportunities for research on aging in basic biomedical science, social and behavioral sciences, clinical medicine, health services delivery, and biomedical ethics. The group sought to document and outline recent exciting developments in aging research, to set clear priorities for future research, and to estimate the resources required to enter into such an agenda for the nation's research scientists. A major thrust of this report examines processes of discovery and treatment that will improve *function* in the elderly population, without denying the importance of research whose focus is restricted to investigation of disease.

This study includes three stages. During the first stage, commissioned papers were developed, and a detailed bibliometric analysis of aging research was conducted. In the second stage the expert committee met several times to deliberate and prepare this report. For the report the committee sought input from a wide variety of agencies and organizations interested in or supportive of research on aging. The committee has reviewed the current funding inventory as well as the results of the bibliometric analysis. Liaison teams comprised of experts in the areas of biomedical, clinical, behavioral and social, and health services delivery research were charged with developing a prioritized list of special research opportunities and needs for the next 10 to 20 years. In addition, two national authorities in the field of biomedical ethics provided information about research needs in this area.

After reviewing and discussing the reports of the liaison teams and ethicists, the committee sculpted the overall research agenda. Special efforts were made to identify a clear set of priorities that reflected both promising and neglected areas of inquiry rather than merely providing a compendium of current or needed research in aging. Particular attention was paid to identification of areas of age-related research that should be supported by the federal government as well as those areas that deserve the attention and support of private foundations.

In the final or third stage of this effort, the IOM will work to bring the report to the attention of policymakers, the scientific community, and the general public.

The National Research Agenda on Aging builds on two previous documents developed by the NIA—a 1978 report, *Our Future Selves*, and a 1982 report, *Toward an Independent Old Age: A National Plan for Research on Aging*—which, taken together, provided an excellent point of departure for the present work. Another major resource was an inventory of federal research on aging, compiled in 1982 by a task force of the U.S. Department of Health and Human Services.

The IOM is uniquely qualified to develop this national research agenda on aging. Since its inception in 1970 as a branch of the National Academy of Sciences, one aspect of the IOM's mission has been to advance and protect the health of the public. Several efforts over the past decade have established the IOM as an important national resource for the analysis of health policy issues on aging. These reports include (1) *Aging in Medical Education* (1978), (2) *Health in an Older Society* (1985), (3) *Improving the Quality of Care in Nursing Homes* (1986), (4) *Productive Roles in an Older Society* (1986), (5) *Academic Geriatrics for the Year 2000* (1987), and (6) *The Social and Built Environment in an Older Society* (1988). The effectiveness of the IOM in influencing public policy is evidenced by the fact that the reports on nursing home care and on academic geriatrics recently led to adoption of federal legislation implementing IOM recommendations.

In concluding this analysis, the IOM drew on special strengths: the voluntary nature of the experts' participation in IOM efforts; IOM's reputation for objectivity; its interdisciplinary membership and perspectives; its collaborative interaction with other components of the National Academy of Sciences; and the thorough, robust, and careful review of all reports prior to their release.

The committee applied important criteria in prioritizing research areas during its deliberations on this project. To qualify for consideration for a prominent place on the National Research Agenda on Aging, the area of research must have fallen within one or more of the following general categories:

- General aging processes: Such studies focus on theories of aging (diverse mechanisms of aging processes, including genetic and environmental components and their interaction) functioning at the molecular, cellular, organ, organ system, organism, and wider psychosocial and sociocultural levels.
- Important age-related disease: This might be a disease occurring predominantly in aged populations or a disease found across the lifespan whose occurrence in old age is associated with specific alterations in its presentation, course, or sequelae (e.g., diabetes).
- Factors influencing age-disease interaction: A number of lifestyle and socioenvironmental factors have an important influence both on the emergence of disease in aging populations and on the mode of presentation by the older victims of disease.

- The functional capacity of the elderly: The enormous importance of functional capacity as a determinant of the care needs of older persons dictates a special focus on those factors that limit the activity of elderly individuals and impair their independence.
- The feasibility and timeliness of the research: The committee was sensitive to the fact that "pie-in-the-sky" research agendas must be avoided in favor of feasible research, that is, research whose "time has come." Similarly, an overly conservative research strategy must be avoided since major breakthroughs often are made when special, timely windows of opportunity present themselves. Other criteria for areas of research included their potential to (1) enhance research in other areas, as in basic studies on proliferative capacity of cells to improve studies on carcinogenesis; (2) reduce morbidity and mortality in older persons; (3) decrease costs of care; (4) increase knowledge of behavioral and social factors in health and disease; and (5) improve pharmacological treatment of patients.

During its deliberations, the committee identified 11 emerging and overarching themes relevant to the entire spectrum of research on aging. These provide a context for the consideration of the individual research areas discussed in the ensuing chapters.

1. An interdisciplinary approach to studies of aging: Many of the most intractable problems in aging research, particularly in the clinical, social, and behavioral arenas, can be properly addressed only by an interdisciplinary group. Interdisciplinary work is costly and often more difficult to design and conduct successfully than unidisciplinary research.
2. Increasing the body of data based on longitudinal studies: Much of the available information on aging is based on crosscohort studies that compare individuals of different age groups. Although these studies provide some perspective on the effects of aging, they are subject to a number of non-age-related influences, such as secular and cohort effects. Longitudinal studies in which serial

prospective measures are made on the same individuals over time provide a much more robust approach to gerontologic and geriatric research. Longitudinal studies, although inherently more time-consuming and expensive than crosscohort studies, are relevant to all domains of research—from basic science through clinical investigation to health services research.

3. Developing research resources—databases, well-defined study populations, cell lines, animal colonies, and animal models for research on aging: Such research resources often can be used by many different investigators conducting various studies at the same time. Federal and, in some cases, foundation funds can be brought to bear to establish such resources. Increasing attention must be given to the types of resources needed to promote the development of aging research to its next level of accomplishment.
4. Areas of neglect—gender, race, cultural background, and ethnicity: Substantial increases in research on aging in members of different racial, cultural, and ethnic groups are needed to clarify the mechanisms underlying differences in the presentation, course, and sequelae of a variety of geriatric-related disorders. Additionally, gender has been shown to have a major influence on biological as well as social and behavioral aspects of aging.
5. Importance of a life-course perspective: Substantial disability and health care expenditures in old age are rooted in the lifestyle, environmental, and other psychosocial factors that begin during youth or middle age. A developmental life-course perspective should be increasingly incorporated into aging research.
6. Emphasis on studying basic mechanisms of aging: Throughout its first phase of development, most gerontologic and geriatric research was descriptive in nature. Studies of the mechanisms of age-related alterations that build on this information and the expanded application of the new and powerful investigative techniques now available are needed.

7. Specialized health promotion and disease prevention research in older populations: A revolutionary increase in life expectancy has occurred already. A corresponding increase in *active* life expectancy or health span should be the focus of much of the next phase of aging research. Such an effort must avoid the simple generalization of health promotion/disease prevention research in middle age to older populations. Instead, it should take into account the special physiological characteristics of the elderly and should focus in particular on such disorders as dementia, incontinence, and falls, which are particularly common and disabling in elderly populations.
8. Studies of the role of genetics, social, and environmental factors as modifiers of aging: The effects of the aging process itself have been exaggerated; the modifying effects of the individual's genetic background, nutritional status, exercise, personal habits, and psychosocial factors have been underestimated.
9. The psychological and sociological context of the individual: Interdisciplinary approaches that consider both the influence of the individual's behavior and environment and social interactions and support systems will be crucial. Equally important is the need to explain specific age-related findings.
10. Expansion of health services delivery research: Many of the issues discussed in this section (e.g., the study of the effectiveness of interventions and the investigation of financing of care) may be implemented by the techniques and mechanisms special to the field of health services delivery. This discipline often serves as the interface between scientific application and political policy decisions affecting the delivery of health care to millions of Americans. Health services delivery research can provide objective information to policymakers, assisting them in developing their programs on a more scientific basis and providing tools for analyzing the usefulness of these programs.

11. Ethical considerations in the care of the elderly as a research focus: Previous agendas for research on aging have seriously neglected ethical issues, such as the rationing of health care and the provision of care to and the conduct of research on demented and irreversibly ill older persons, especially with regard to the application of life- sustaining technologies.

One of the major scientific challenges for the United States over the next decade will be to enlarge our research capacity in aging so that we gain the knowledge necessary to deal with the health, social, and psychological needs of a rapidly aging society. This plan provides a prioritized and rational blueprint for that research development.

BASIC BIOMEDICAL RESEARCH

The term *aging* can refer to changes taking place during any stage of life. Some of these changes are benign, with no evident adverse effects; others are eventually harmful. The dysfunctional changes associated with accelerated mortality rates during the later phase of adult life are collectively called senescence. Senescence is associated both with diseases of specific organs and tissues and with diffuse disorders that defy conventional classification. Within the vast scope of these phenomena of aging, many of the problems of senescence can be explored scientifically. Continued research not only will yield new interventions into specific disorders of senescence but also will lead to the discovery of the nature of many basic biological mechanisms that change during the aging process.

A number of puzzling questions about aging have attracted the interest of biomedical researchers. Why, for example, do humans differ so individually in aging changes of bone, brain, and heart? Here we see how genetic and epigenetic factors produce different responses to environmental influences throughout the lifespan. Because the environment begins to influence the organism long before birth, aging represents an array of processes that develop over the entire lifespan. Thus, environmental influences on genetic and epigenetic characteristics of an individual may give rise to preconditions of disability and disability itself long before senescence is manifested. The potential for reversing or delaying disability depends on the still unknown nature and extent of fundamental changes in gene expression through the aging process. Careful study of environmental-genetic interactions throughout life is required.

Clearly, aging is a unique frontier of the life sciences that requires examination by a wide array of scientific disciplines. Research in aging will clarify the basis of many specific disorders of human aging and, in so doing, will add to the proud advances that have eliminated so many diseases of children and younger adults. Moreover, the ability to understand aging requires a powerful intellectual synthesis of diverse research areas that presently stand apart from each other because of their historical focus on specific organs and diseases.

Fundamental information about the basic mechanisms of aging still is woefully inadequate. Critically necessary is aging research in the basic disciplines of biology: genetics, biochemistry, cell biology, neurobiology, developmental biology, and others. The history of science amply demonstrates that major advances often come from serendipitous discoveries.

Thus, in any agenda designed to establish priorities and to estimate the resources for aging research, it is of utmost importance that investigator-initiated research be protected. This should provide the core from which many major concepts and discoveries will emerge.

Research on basic cellular functions has brought many new insights into the biological mechanisms of aging (Röhme, 1981; Stanulis-Praeger, 1981) and research gives ample reason for optimism.

Equally, new, improved, and expanded model systems for the study of aging have been uncovered; these include rodents from food-restricted colonies with enhanced lifespans and delayed, reduced, or absent age-related pathologies (Masoro, 1988); methods for the creation of strains of laboratory mice with a wide array of targeted mutations (selection of substrains with specific genetic characteristics); breeding lines of unusually long-lived insects from genetically heterogeneous stocks (Dice and Golf, 1987); and identification of a single gene mutation that increases lifespan in the nematode *Caenorhabditis elegans*. Progress has been made in the development of chemically defined media for cultivation of normal diploid somatic cells to facilitate analysis of mechanisms of clonal senescence and cellular repair.

Major conceptual and methodological advances in the techniques of molecular biology, especially in molecular genetics, are leading to an increased understanding of gene expression at the molecular level. Genetic maps are reaching high resolutions (markers at about one million nucleotide base pairs), with the potential for mapping and cloning the dominant genes responsible for such various age-related disorders as familial Alzheimer's disease (St. George-Hyslop et al., 1987). Finally, scientists are increasingly aware of the intellectual challenges posed by biogerontology as a major uncharted frontier of biology.

The committee believes that elucidation of the biological mechanisms of aging is an achievable goal. The major recommendations that follow focus attention on a broad array of important areas of investigation that are both feasible and critical to our knowledge of aging and its basic processes. Three main criteria were utilized in the selection of the research priorities in the field of basic biomedical investigation:

1. Feasibility of the research in the context of current advances in the biological sciences: Enough is known to permit immediate expansion of existing efforts in these areas and to plan for and develop the resources needed during the next decade. Expansion of these research areas will have an immediate impact on the understanding of a number of aspects of the aging process.
2. Importance of these research areas insofar as potential findings can be applied to the treatment of major disabilities of later life: In the examples discussed, the committee believes that biomedical research eventually will lead to prevention of these disorders.
3. Potential of the research areas selected to catalyze a cascade of productivity in many other areas of health research, including areas outside of aging that share technology or concepts with gerontology.

RESEARCH PRIORITIES

- The committee proposes a major new initiative to achieve understanding of the basis for the pervasive disturbances in the regulation of proliferative homeostasis that accompany aging in essentially all animals, including humans.

These disturbances in proliferative homeostasis, a fundamental process responsible for appropriate replacement of cells that have become lost either because of exposure to toxins or because of endogenous physiological processes, play major roles in cellular regeneration and repair and in the genesis of several of the most important age-related human disorders: cancer, atherosclerosis, osteoarthritis, benign prostatic hyperplasia, and altered immune function (Martin, 1979). This new major national initiative should be comparable in emphasis and scope to the NIA's highest research priority— Alzheimer's disease and the neurobiology of aging.

- The second priority research recommendation is that basic research in the neurosciences (including the peripheral and central nervous systems) and the current special research initiatives on Alzheimer's disease should be continued and expanded.

Funding allocations for the foregoing should supplement, not supplant, existing research support and should not detract from other areas of investigation. Fundamental studies on aging and the regulation of gene expression and macromolecular syntheses, postsynthetic modifications of proteins and protein degradation, membrane changes, and other fundamental research approaches are essential to the studies envisioned above and to all facets of biomedical exploration. To achieve the above-noted research goals, major new research resources are required for direct support of research projects, additional training programs, expansion of current centers devoted to studies on aging, and enlargement of the current infrastructure for basic biomedical scientific exploration. These are described later in this chapter.

ADDITIONAL RESEARCH OPPORTUNITIES

The identification of two major research recommendations in no way is intended to detract from the importance of other areas of biogerontological research. Some promising research directions not included in the major recommendations also are worthy of encouragement and support. A brief discussion of some of these important research directions follows. Although one major emphasis of this report is on proliferating cells, the role of nondividing cell types in the process of aging should be explored further. These studies should focus on postmitotic cells, such as neurons and cardiac and skeletal muscle cells, as well as on conditionally proliferating cells, such as hepatocytes (Martin, 1977), and will provide important information about aging mechanisms that have been neglected. Presently, little or no information is available about repair and regeneration in these cells. For example, does macromolecular turnover change with age? How are cell surface properties altered? Do signal transduction mechanisms change?

Another area of emphasis is that of systems physiology. Most gerontologists agree that aging is a multifactorial process, involving many cells, tissues, and organ types. Research must be carried out on the major integrative systems in physiology. The effects of aging on the endocrine/neuroendocrine and immune systems (Finch et al., 1985; Miller, 1990) must be understood if we are to understand the organism's aging phenotype. Dietary restriction as a modulator of lifespan in rats and mice represents an important probe for understanding aging changes (Masoro, 1988). Similarly, regulation of reproductive physiology and of the developmental process in general represents a potentially important analogue for the aging process and should be emphasized (Finch et al., 1985). The fact that cells, tissues, and organs do not age at the same rate among species or even among individuals within a species poses a key complicating factor for these studies. Hepatocytes, for example, may be functionally youthful in an individual whose nervous system or cardiac function is seriously impaired as a result of aging or disease.

The use of lifespan as an end point for aging has further hampered the interpretation of numerous studies. The field sorely needs measures of senescence based on functional capacity. Thus, descriptive studies of each system, evaluated in relation to overall functional competence and mortality risk, are critical to the development of meaningful biomarkers for aging, and for the identification of dysfunctional aging (senescence). Such biomarkers would increase scientific understanding of the factors that influence the rate of aging along the continuum of biological change and would contribute to the development of interventions that might delay or reverse dysfunctional aging (senescence). The concept of biomarkers applies at many levels, from cellular biology to the more complex interactions that are the object of scientific study in clinical, behavioral and social, and other areas of health care research (Sprott and Baker, 1988). Abundant evidence is available of the accumulation of abnormal proteins during the course of aging and in the development of age-related diseases, such as the neurofibrillary tangles and beta-amyloid of Alzheimer's disease (Stadtman, 1988). Moreover, many tissues acquire inactive enzymes during aging (Dice and Golf, 1987). A major question concerns the pathogenesis of such abnormal proteins. The general question of altered gene expression in aging is of the utmost urgency if we are to understand the biological bases of aging, and the role of environmental influences, such as free radicals, radiation, and various toxicants, raises an additional and pressing set of questions about how the environment may influence aging (Ames et al., 1985).

While some research is focusing on various types of molecular damage, studies on the mechanisms of selective changes in gene activity should receive special emphasis. The powerful tools of recombinant DNA genetics are being applied to these questions. For example, the new technique of polymerase chain reaction can be used to ascertain if new genes are being activated in individual cells of different tissues. It is possible that alteration in the activity of a limited number of genes is a common denominator for many cellular changes in metabolism during aging. In effect, many of the same kinds of questions about selective changes in gene activity during development of the organism arise when considering changes resultant from the aging process.

The development and characterization of various model systems must receive more attention. The study of comparative genetics and lifespan is a challenging and potentially productive area of research for learning more about aging. New model systems should be developed based on their usefulness in responding to specific biological questions and on their value for comparative genetics. Models are useful insofar as they can be used to answer specific questions about biologic phenomena or mechanisms.

Other areas for productive research include the development of optimal methods for the isolation and cryopreservation of tissue biopsies and purified cell populations; extension of current epidemiological studies on the interaction among environment, genetic factors, and aging; and the application of life- parameter techniques to animals of varying lifespans.

RESOURCE RECOMMENDATIONS

The following summarizes the major resources that the committee believes will be needed to meet the foregoing goals. A discussion of the funding required to develop these resources appears in the section on funding in the Executive Summary and Recommendations for Funding.

Funding of Research Proposals

The NIH funds about one in four approved research proposals. The committee believes that many worthy research questions are not being examined because of lack of funding and that many promising lines of research are being abandoned or delayed. Therefore, the most important way in which the agenda on basic biomedical research (and other areas in the study of aging) can be implemented is to increase the rate of funding for approved research projects on aging from its present level of one in four to one in two. Clearly, research proposals not only must be approved but also must be of high quality in order to be supported by the additional funding.

Materials

Animal Models

More species are needed, including primates and other long-lived animals, that resemble humans more closely than those short-lived rodent models having condensed reproductive schedules. There is also a need for expanded support of programs within the NIA and other institutions to make available to researchers aging cohorts of animals of contrasting maximum lifespan potentials.

Cell Line and Tissue Samples

Increased effort should be directed toward making available cryopreserved cells and tissues. Improved access to these resources has already had an important effect on the progress of tumor biology. A new initiative is required to develop the technology of cryopreservation, banking, characterization, and distribution of postreplicative cells from donors of varying ages and genotypes. Samples of both of these cell and tissue types should be derived from different human populations according to age, sex, and state of disease (including, but not limited to, Type II diabetes mellitus and Alzheimer's disease).

Cell lines should be developed from more species, including long-lived avian species such as the Japanese quail, to evaluate lifespan relationships and propensity for spontaneous transformation. Finally, cell lines should be obtained for study from species having different lifespans and from human donors with certain progeroid syndromes (Martin, 1979).

Further Development of Existing Longitudinal Studies

Although this chapter focuses on fundamental biological processes of aging, this effort is directed primarily at the problems of human aging. Therefore, the committee stresses the resource importance of epidemiological studies of how humans age. Efforts should be made to increase the accessibility of archival information, especially the data from previous major longitudinal studies. These studies include information about childhood and adolescence that may form the basis for future gerontological research (Verdonik and Sherrod, 1984) and a current NIA crosscultural study of Alzheimer's disease that compares Japanese-Japanese with Japanese-Americans. Also, data on mortality rates and incidence of disease or dysfunctions of different human and animal populations should be made available.

Ten additional Centers of Excellence in Research and Education in Geriatrics and Gerontology (Claude Pepper Centers) should be established to maintain the foregoing resources. Also to be developed is resource support to perform high-technology tasks, such as specialized laboratories for cell biology, that are fundamental to a wide range of aging research. Three such laboratories are recommended to serve as regional infrastructure support for basic biomedical studies on aging.

There is evidence that the absence of sufficient laboratory space has been a major impediment to more rapid progress in Alzheimer's disease research. Congress can address this problem, in part, by appropriating construction funds already authorized by legislation creating Alzheimer's Disease Research Centers. In addition, we need to produce at least 200 more well-trained research scientists per year with commitments to research on aging. To the aggressive pursuit of young researchers by existing training programs should be added new programs designed specifically to recruit scientists with established reputations in other fields.

In view of the special nature of gerontological research, the committee strongly recommends that mechanisms for long-term research and research training support (5 to 7 years) become more widely available. New funding for research is required to implement these initiatives. Further sums will be needed for training, infrastructure support, and construction and renovation to house the research resources in the field.

In addition, the basic biomedical research program will share in the development of additional multidisciplinary centers for research and education (Claude Pepper Centers) and in the support of infrastructure resources (e.g., computer capability; data banks, including population studies; and library support).

With additional support for research grants, infrastructure, and centers, a significant increase should follow both in the number of investigators who turn their attention to the phenomena of aging and in a stepped-up intensity in the nature of the research currently under way. Increased support for research careers in the study of aging should also lead to greater enrollment of young professionals in those training programs now undersubscribed. Both cancer biologists and gerontologists are now well aware of the close interrelationships in the regulation of senescence and neoplasia.

The expanded effort proposed in this report can be expected to lead to an increased understanding of the nature of cell regulation and how changes in this process lead to those diseases of aging characterized by aberrant proliferation, such as cancer, atherosclerosis, osteoarthritis, and prostatic hypertrophy. In addition, understanding the genetic

and epigenetic regulation of functional capacity as manifested at the cellular level will provide models to understand senescence and its regulation in a wide variety of cell and tissue types. Insight into these regulatory factors will yield substantial dividends in terms of the prevention of and therapy for the illnesses and disabilities of older individuals.

CROSSCUTTING ISSUES

Biomedical research has only recently begun to examine the role of gender, race, and ethnic background and the relevance of those issues to altered trajectories of aging. The gender differential in lifespan is one fundamental issue that needs clarification. In developed countries, for example, females enjoy a substantially greater average lifespan. Although a major reason for this difference perhaps can be attributed to lifestyle differences (smoking, alcohol, violence, and fat ingestion), other reasons for the female lifespan advantage are obscure. For other mammalian species it is still not clear whether a consistent gender differential in lifespan really exists. In any case, sociobehavioral factors perhaps are crucial in these effects and in the gender differential in the incidence of many age-related disorders.

Another crosscutting issue is that of ethics. What are the ethical considerations in doing basic biomedical research? What about the ethical questions attendant upon genetic engineering? In 1990 human gene transplants became a reality. What are the ethics concerning those gene transplants that might prevent the development of age-related dysfunctions but that might as well have adverse effects on younger individuals? Finally, there is and will continue to be a need for regularly updated interdisciplinary education, which should be developed at several levels of sophistication and targeted for a range of professional specialization in health caregiving and administration.

CLINICAL RESEARCH

The rapid growth of clinical research in geriatric medicine over the past decade permits a broad-based attack on myriad important and previously neglected research questions across several research areas. These include investigations of age-dependent changes in an array of physiologic systems, pathophysiology and management of a number of diseases relevant to aging populations, and characterization of the problems of frail, elderly people, including those in long-term care facilities.

Building on the impressive though unsystematic database that has been developed, a consensus is emerging regarding the most important lines of future research inquiry. That consensus, sharpened by the systematic discussions involved in preparing this National Research Agenda, focuses on two key areas that can serve as the basis for specific research questions and proposals: (1) research into the causes, prevention, management, and rehabilitation of disability in older persons, including a focus on geriatric syndromes; and (2) studies of age-dependent physiologic change and specific pathologic entities.

Recommendations for specific studies within these two major topic areas overlap significantly, reflecting both the multifactorial nature of disease and disability in old age and the absence of clear, common pathways of declining function in older people.

Such overlap also is inherent in the area of clinical research since the field ranges from molecular biology and genetics to epidemiology, and draws on information supplied by basic biomedical research, health services studies, and the behavioral and social sciences. Clinical research shares many techniques and subjects with other areas of investigation but has a special focus on the connections between organic illness and functional status and on specific interventions to treat or prevent organic diseases. Several recent Institute of Medicine (IOM) studies on research into the effectiveness and outcomes of health care have described the potential for interaction between clinical research and other areas of investigation in the study of older patients (IOM, 1989, 1990a).

In addition to overlapping with other areas of research, the clinical research priorities described below share a common opportunity for investigation —study of the value of rehabilitative care in clinical outcomes, in cost benefits, and in cost effectiveness.

RESEARCH PRIORITIES

- The first priority is research into the causes, prevention, management, and rehabilitation of functional disability in elderly persons. The overwhelming importance of functional capacity as a determinant of older persons' needs dictates a special focus on factors that limit the independence of elderly individuals. Increasingly, the clinical geriatric research community recognizes and agrees that it is well positioned to make significant short-term advances in the understanding of the physiologic and pathophysiologic mechanisms underlying the dependency and frailty that characterize the final stages of life.

Although at any given time, a majority of older persons are free of significant disability, the rate of disability rises with age. The National Health Interview Survey (National Center for Health Statistics, 1985) reported that 23 percent of all elderly persons are unable to perform at least one of the activities of daily living (ADLs).

These ADLs include bathing, dressing, grooming, going to the toilet, being able to move from bed to chair, being continent, and being able to feed oneself. About half of those 85 years of age and older need assistance to perform one or more ADLs or to carry out one or more instrumental activities of daily living (IADLs). These include shopping, using public transportation, cooking, using the telephone, housekeeping, and managing finances (Dawson et al., 1987).

In order for an older person to be independent, he or she must be able to perform these activities. Clearly, with increasing inability to per form ADLs and IADLs, one's dependency on others for care increases. As many as one to one-and-a-half million elderly individuals residing in the community or in institutions are unable to carry out five or more ADLs (Branch et al., 1984). However, recent studies indicate that such deficits are not static and may indeed respond to rehabilitation (Williams, 1988).

Within the general area of functional disability research, two general areas stand out: studies of health promotion and disease prevention in older persons and studies of specific geriatric syndromes.

Health Promotion and Disease Prevention in Old Age

The mechanisms and diseases that lead to age-related increases in disability have not been elucidated. With few exceptions, research in health promotion and disease prevention has not focused on older cohorts.

Although substantial data on the relationship of traditional cardiovascular risk factors to the subsequent incidence of cardiovascular disease in older men and women are emerging from the aging Framingham cohorts, few data are available to indicate whether intervention in these risk factors affects subsequent rates of cardiovascular disease, including stroke, heart attacks, and congestive heart failure.

Some limited data exist in the area of diastolic hypertension, and a major multicenter study is under way to evaluate the contributions of isolated elevations in systolic blood pressure to these end points, particularly stroke. However, few studies adequately address remediable risk factors for other important geriatric disorders, including osteoporosis, dementia, disability, and total mortality.

Too often, investigators have generalized findings from middleaged adults to older individuals without recognizing the major effects of the substantially altered physiologic substrates with advancing age. No a priori assumption can be made that risk factors for important diseases in middle age carry the same relative or the same attributable risk in old age. From a lifespan perspective, it is important to note that many prevention issues pertinent to older people should first be addressed to young adults (IOM, 1990a,b).

While a substantial number of specific research areas have been identified, two are particularly salient:

1. Research should be conducted on the interacting effects of age, lifestyle factors, and disease on the disability status of older persons.

2. Studies should be undertaken to determine the effectiveness of various disease prevention strategies or practices in older persons.

Studies of Geriatric Syndromes

While functional disability may be the final common pathway of frailty for many older individuals, the specific clinical manifestations of disability vary among individuals. In many patients functional dependence is associated with the emergence of one or several specific geriatric clinical syndromes that may share some basic underlying pathophysiologic mechanisms. Geriatric syndromes represent a cluster of symptoms, conditions, and disabilities resulting in a variety of physiologic changes, pathologic conditions, comorbid conditions, and environmental challenges.

These disorders are quite common in persons over age 75 and result in substantial disability, morbidity, and decreasdd quality of life. A number of these syndromes—including failure to thrive; impaired postural stability, strength, and mobility; mismanagement of medications; urinary incontinence; and delirium—have been neglected for some time but are particularly ripe for major research advances.

Failure to Thrive (Inanition)

This is a syndrome of weight loss, decreased appetite and poor nutrition, and inactivity, often accompanied by dehydration, depressive symptoms, impaired immune function, and low cholesterol.

Occurring in both acute and chronic forms, failure to thrive leads to impaired functional status, morbidity from infection, pressure sores, and, ultimately, increased mortality. Failure to thrive also leads to increases in institutionalization and health care costs. One example of promising research into this syndrome is the conduct of crosscohort and longitudinal studies to determine the prevalence and the underlying risk factors precipitating failure to thrive. Another important example is the study of interventions to prevent or treat such consequences of failure to thrive as pressure sores and depression.

Impaired Postural Stability, Strength, and Mobility

This is a syndrome that includes the common problems of falls, dizziness, syncope, fractures, muscle weakness, and impaired mobility. These problems tend to occur increasingly in persons over the age of 75 and result in significant functional disability and dependence. Each year in the United States there are over 200,000 hip fractures—the majority of which are precipitated by falls—in persons over the age of 65 (Cummings et al., 1985). One specific research recommendation in this area is to determine the pathophysiologic mechanisms and other factors underlying recurrent falls and mobility impairment, including gait disorder and muscle atrophy. Further study is needed to explore effects of pharmacologic agents and other therapy (graded exercise) on improving or impairing some of these functions.

Mismanagement of Medications

The management of multiple medical conditions in older patients often involves several medications. Polypharmacy, defined as the taking of three or more medications on a regular basis, is found in a third of persons over the age of 65 (Nolan and O'Malley, 1988). Recent studies indicate that polypharmacy is a significant cause of morbidity and hospitalization among elderly patients (Williamson and Chopin, 1980; Ives et al., 1987; Grymonpre et al., 1988). Many of the geriatric syndromes described in this report occur as side effects of prescription medications. Yet, at the same time, many older persons' easily treated conditions are not managed aggressively enough with the appropriate pharmacologic agents.

This mismanagement of medications compromises independence and quality of life, and it increases the costs of health care. More studies of the fundamental pharmacodynamics of various medications in older individuals would forward clinical care in this area. Additionally, opportunity exists for excellent ground-breaking interdisciplinary investigation of medication use and effects that link basic clinical and behavioral science approaches. This might include gene delivery systems (cells carrying engineered genes), specific enzymeblocking agents, and monoclonal antibodies.

Urinary Incontinence

Urinary incontinence, although it varies in intensity and in the degree to which its victims are disabled, is a highly prevalent, costly, morbid, and seriously neglected problem in elderly persons. The prevalence of urinary incontinence is approximately 30 percent in community-dwelling older persons (Diokno et al., 1986) and reaches 50 percent for those in nursing homes (Ouslander et al., 1982). Costs of incontinence are measured not only in terms of the financial burden of its management (e.g., linen changes; use of absorbent pads; and use of diapers, condoms, and catheters) and its clinical morbidity, but also in psychosocial terms. Our knowledge regarding the mechanisms underlying urinary incontinence and the appropriate approach to its control is rudimentary. There is a great need for clinical studies to determine the neuroanatomic and neurophysiologic basis of urinary incontinence and the efficacy and risk of treatment (e.g., drugs, surgery, training, and elimination of risk factors). Finally, there is an urgent need to develop new, effective, and safe treatments—preventive, ameliorative, or curative —for incontinence.

Delirium

Delirium, the development of acute confusional states, is a global disturbance of cognitive function accompanied by an alteration, and often fluctuation, in the level of attentiveness and, usually, consciousness. Delirium frequently is overlooked or misdiagnosed, and often mistaken for dementia. Approximately one-third of hospitalized elderly individuals develop an acute confusional state, markedly complicating their hospital course and dramatically increasing not only morbidity but also health care costs (Lipowski, 1989). Despite its high prevalence and tremendous morbidity and costs, little is known about the underlying mechanisms of or effective interventions for delirium.

- The second research priority involves studies of the interaction of age-dependent physiological changes and important diseases in old age. The scientific and medical community needs to evaluate the interaction of age-related physiologic changes with diseases that either occur predominantly in aged populations or present different

> symptoms, course, or sequelae when occurring in the aged. This area of research is largely interdisciplinary since a number of lifestyle and socioenvironmental factors have an important influence on the emergence of disease in aging populations. Research on the interaction of aging and disease processes provides a critical link to issues of functional capacity since each major disease category mentioned in this report imposes a large burden on the aging population; prevention of or more effective intervention into these disorders might well reduce the onset of disability.

To effect the study of the interaction of aging in disease, research knowledge of biologic and physiologic processes from many levels must be integrated. Crucial to this effort is the appreciation that research gains in the areas of both normal and disease states act synergistically to increase our understanding of pathophysiologic mechanisms. This approach has been particularly well demonstrated in the use of molecular biological tools that delve into genetic regulatory processes that underlie the control of both normal and abnormal molecular and cellular processes. To gain an understanding of age-related pathophysiology, the medical and scientific community must exploit the most fundamental techniques of molecular and cell biology and apply them to the investigation of normal human aging. There is reason to be optimistic that a better-understood and better-mapped human genome, coupled with the application of now- standard techniques of reverse genetics, will allow the genetic contributions to many of these diseases to be defined specifically. This development, plus the extension to actual gene therapy of current techniques to introduce cells carrying engineered genes, may allow us to prevent, cure, or otherwise substantially modify many of these diseases.

Cardiovascular Disease

Cardiovascular disorders are a major target, for cardiac disease is the number one cause of death in persons over the age of 65 and cerebral vascular disease is the third most common cause of death in elderly persons (National Center for Health Statistics, 1989). Most of these disorders are due to atherosclerosis: excessive thickening and narrowing of coronary, cerebral, and peripheral arteries. Unfortunately, inadequate data address these problems because, to date, elderly persons have not been represented adequately in clinical studies of cardiovascular therapies. Thus, increased research on cardiovascular disorders among older persons is needed. Emphasis should be placed (1) on understanding the genetic as well as the environmental risk factors and the molecular basis of atherosclerosis, and (2) on developing standard and recombinant-based drug technologies as well as new approaches to prevent (in younger adults) and treat hypertension, lipid disorder, and the thrombotic complications of atherosclerosis. The role of nutrition, and the interaction of genetic and nutritional factors in the cause and prevention of these diseases, are important areas for future research.

Dementia, Pseudodementia, and Psychiatric Disorders

Recent studies have shown that dementia after age 75 reaches well over 40 percent; an overwhelming proportion of these patients have senile dementia of the Alzheimer's type (Hagnell et al., 1983; Sayetta, 1986). If dementia were cited on the death certificates of persons in whom it was a primary underlying disease, it would be the fourth or fifth leading cause of death in the United States (Weiler, 1987). Furthermore, in this country dementia now is the leading cause of institutionalization (Von Vostrand, 1979). Affective disorders are underreported in older persons even though they account for substantial disability in this age group (Finlayson and Martin, 1982). Although there are many outstanding research opportunities in this area, the greatest need is for studies linking basic and clinical approaches to the pathophysiology, cause, prevention, and treatment of Alzheimer's disease and other dementias.

Excellent opportunities are available for basic molecular and cell biologic studies to determine the mechanisms and causes of central nervous system cell death in the aged, particularly the role of growth factors in central nervous system changes associated with both normal aging and injury. These studies should include assessment of cellular and molecular markers and new approaches to imaging and neurotransmitter mapping of the nervous system. In addition, clinical studies that focus on management strategies for demented older persons have major interrelations with social, behavioral, and health services research agendas.

Musculoskeletal Disorders

Musculoskeletal disorders rank second to circulatory system diseases as a cause of both disability and health care costs in the United States. These conditions rank second in frequency as a reason for visits to physicians and third in frequency of hospitalization. Osteoarthritis is the most prevalent joint condition after age 65 (National Center for Health Statistics, 1989), and ranks second (after cardiovascular disease) in producing severe disability in those over age 60. Osteoporosis is also extremely common; annually, over one million osteoporosis-related fractures occur in the United States (Resnick and Greenspan, 1989). Fractures of the vertebral bodies and the hip are the two most common types of fracture in elderly persons, and both cause significant morbidity of those suffering from hip fracture, 50 percent never walk again (Melton and Riggs, 1983). Resultant costs from these disorders total more than $65 billion annually.

The application of restriction fragment-length polymorphism and other techniques of reverse genetics will help to provide the tools necessary to define the genetic basis of osteoarthritis and osteoporosis in many patients suffering from these common disorders. Such approaches should receive a high priority, particularly if current techniques of genetic engineering can be extended to gene replacement.

Infectious Disease and Immunosenescence

Infections, especially pneumonia and those of the urinary tract, are a major cause of disability and mortality in old persons. However, the exact extent of disability from infectious disorders in older people is largely unknown. Knowledge is incomplete about the prevention and treatment of infection in this population. Specific research opportunities include studies of the basic mechanisms of immune deficiency in older individuals, including changes at the cellular level involving T-cell subsets, natural killer cells, macrophages, and receptor-mediated responses. Vaccine development needs to take into account the special characteristics of the elderly population. Vaccines that can bypass the need for immune T cells and vaccines made with live attenuated organisms with antigen attached to a rigid backbone may be more effective in old people. These efforts should utilize new technologies such as DNA recombinant vaccines and genetically engineered attenuated viruses. The focus on immunosenescence is also directly related to the next research focus, cancer in older persons, since impaired immunity contributes to cancer development.

Neoplasia

Cancer and its related complications are the number two cause of death in persons over the age of 65 (National Center for Health Statistics, 1989); for elderly persons death due to cancer is the second leading cause of lost years of life. The basis for the age-associated frequency of and disability related to cancer is unknown. In addition, efforts at primary (e.g., prevent onset of disease), secondary (e.g., prevent onset of complications of disease), and tertiary (e.g., treat complications of disease) prevention of cancer have been limited almost entirely to the younger population. For example, little information exists on the efficacy of smoking cessation programs and the efficacy and toxicity of standard chemotherapeutic regimens for those 65 years or older. As with several of the other suggested areas of research identified in this report, there is a clear and substantial overlap between the clinical research related to neoplasia and the focus on aging and cell proliferation suggested in the basic biomedical component of this research agenda. Reverse genetics, recombinant DNA products, and delivery systems using gene trans fer approaches should be particularly valuable in this area. Interactions between environmental (e.g., carcinogens and nutritional factors) and genetic factors in the cause and prevention of cancer warrant intensive study.

Disorders of Metabolism and Homeostasis

Strong epidemiologic evidence indicates that metabolic dysfunctions have significant impact on the pathogenesis of numerous disorders affecting the health of older individuals. Metabolic problems may precede the development of clinical manifestations of disease, accelerating the progression of atherosclerosis, hypertension, neuropathy, and musculoskeletal disease. Disorders of metabolism and homeostasis influence the onset and severity of most of the alterations described elsewhere in this report. The metabolic disorders considered in generating these recommendations include diabetes mellitus, altered muscle metabolism, altered bone mineral metabolism, and lipid disorders. Intensive studies should be undertaken to clarify the pathophysiology and genetic basis and to elucidate the long-term complications of these disorders in elderly individuals. Research on relationships among genetic, pathogenic, and nutritional factors will be of great value.

ADDITIONAL RESEARCH OPPORTUNITIES

Although the committee emphasizes the foregoing areas of research for priority consideration, other areas should not be neglected in the research agenda on aging. These additional opportunities include dental and oral problems in aging, especially the prevention and treatment of periodontal disease, cancer of the mouth, and orofacial pain; diagnosis and treatment of neurogenic dysphagia; investigation of hearing impairment, its prevention, and its treatment; study of conditions (particularly glaucoma, cataracts, and macular degeneration) contributing to loss of vision in the aged; and various skin conditions associated with aging (keratomas, skin cancer, and dry skin).

While each of the foregoing research areas leads to detailed specific secondary research objectives, each area also represents the interface of several disciplines. Major opportunities in clinical geriatric research exist in interfaces between physician-scientists and colleagues from the social and behavioral sciences. Increasingly, productive interaction between basic and clinical scientists is resulting in research strategies and protocols that provide the opportunity for application of molecular biological and other basic science techniques to clinical problems and populations.

RESOURCE RECOMMENDATIONS

The ambitious, though appropriate and overdue, clinical research program recommended in this chapter can only be implemented effectively if two major resources become available.

The first is funding to support the specific research projects that flow from the two major thematic areas in this chapter (described in greater detail in the Executive Summary and Recommendations for Funding). Additional funding is required to launch an effective research program in the major consensus areas identified as most deserving of support. There would be little delay in the development of excellent research proposals in these areas since the National Institute on Aging (NIA) and other federal agencies already have a substantial number of approved but unfunded research proposals with excellent priority scores. The additional research funds could be applied to such applications, thus raising the level of support from the current less than one in four (24 percent) to one in two. However, allocation of substantial funds to support research projects will not ensure the successful completion of the clinical research agenda. The second major necessary resource is professional talent. For example, a recent IOM report (IOM, 1987; Rowe et al., 1987) shows a critical need for the development of scientists with a commitment to and appropriate training for research in geriatrics. Both basic biomedical and more clinically oriented investigators are desperately needed in all disciplines related to clinical research in older persons. Two strategies are suggested to attract young health professionals and Ph.D. students into the fields of geriatrics and gerontology.

First, increased funds are needed to establish the ongoing fellowship and postdoctoral programs that will be the lifeblood of the next generation of clinical gerontologists. These funds would permit graduation of an additional 200 fellows and postdoctoral students per year, as well as provide support for 180 junior faculty and midlevel investigators who have recently moved into the fields of geriatrics and gerontology.

A second strategy to develop a cadre of outstanding clinically oriented investigators in geriatrics and gerontology is the addition of at least 10 highly sophisticated Geriatric Research and Training Centers (Claude Pepper Centers) to the current three supported by the NIA. This strategy has been recommended in a separate report of the IOM that focuses on workforce needs in geriatrics. Subsequent to that report, enabling legislation has been drafted and adopted by Congress. These centers would be developed in the setting of currently well-established academic research and clinical programs on aging. A moderate amount of core support should be provided to these centers for continuity of key research personnel and ongoing efforts; however, such centers would be expected to compete for additional research funds on a continuing basis to ensure that their work remains current and is peer reviewed. The primary goal of these centers would be to produce substantial numbers of very well-trained, clinical-research- oriented investigators. To date, three centers have been funded. Ample expertise already exists at enough academic programs to warrant immediate funding of 10 additional centers, followed by funding of further centers over the next several years as the need is demonstrated.

Finally, it is abundantly clear that the facilities and infrastructure available in the United States to support the research enterprise are seriously deficient. Accordingly, the committee recommends additional support for infrastructure and for construction of new research space and renovation of outdated space.

CROSSCUTTING ISSUES

Clinical research in geriatrics and gerontology requires an approach that engages many disciplines. Insights from molecular and cell biology provide a basic biomedical foundation for much clinical investigation, including newer therapies for Alzheimer's disease and the diagnosis of age-associated cancers such as carcinoma of the colon. Behavioral and social research and the insights offered by studies in health care delivery make possible clinical programs to study the responses of older people to illness, and to find better ways of intervening to reduce disability and morbidity among the aged.

Of equal importance in clinical research are crosscutting issues that involve gender differences in response to illness or in the metabolic disposition of drugs as well as racial and ethnic variation in the way older individuals experience illness, both from a patient care perspective and from the vantage of exploring differences in physiological response to disease. An example of the influence of gender differences on outcome of disease is seen in the lower rates of morbidity and mortality in women who have hypertension than in men with hypertension. Racial differences also appear to influence prognosis in some diseases; higher death and complication rates occur in hypertensive black men than in white men.

Another critical crosscutting issue is that of ethics in clinical research. Obtaining informed consent from cognitively impaired elders, using drugs with high risk to benefit ratios in geriatric populations, and resolving conflicts arising from confrontations between medical paternalism and the need for patient autonomy are all examples of the pervasiveness of this issue in clinical geriatric medicine and in geriatric and gerontological research.

3

Behavioral and Social Sciences

Recent behavioral and social research has advanced our understanding of the aging process, the health and well-being of older adults, and the experience of growing older in our society. Three conclusions emerge from a review of this research: (1) sociocultural contexts are an important influence on aging processes; (2) significant variability in aging exists among individuals and between social groups; and (3) skills, behavior, and competence can be modified in old age. Increased length of life, which has dramatically altered the age structure of populations, coupled with changes in social policy and family structure, underscores the need to know more about the nature of aging today and its likely effects on future generations.

The primary goal of research in health and aging must be to compress and diminish the duration of morbidity, disability, and suffering during the extra years provided by increased life expectancy and to enhance both productivity and the quality of life during that time. Because of the complexity of life processes, a considerable portion of this research must be multidisciplinary, longitudinal, and cohort sequential, working within existing knowledge, methodology, and resources.

For example, behavioral and social studies can contribute to the understanding of the response to clinical interventions or can make clear the factors that influence older persons' access to and participation in health care services. Although the traditional approach to research is featured in this review, long-term payoffs from less-established but promising areas of research should not be neglected.

The significance of research discoveries in behavioral and social research on aging can be interpreted in light of three important themes, developed in the next section, bearing on how individuals age, experience aging, and respond to aging: (1) the dynamic interaction of older individuals and sociocultural contexts, (2) differentiation among older individuals and in the aging process itself, and (3) modifiability through interventions to improve the quality of aging. Approaches to the scientific study of sociological and behavioral factors in aging include (1) refinement of measurement and analytic instruments, (2) cooperation and coordination by different disciplines in large-scale investigations using carefully selected, culturally representative panels to be followed longitudinally and cohort sequentially, and (3) application and evaluation of research findings through systematic field studies to test the appropriateness of particular intervention techniques.

JUSTIFICATION AND MAJOR THEMES

Fundamental advances in the study of aging have resulted from the demonstration that factors in the social and physical environment extrinsic to the individual can dramatically alter the course of aging. Thus, the process of aging is highly mutable and susceptible to interventions of a social and behavioral nature. The committee believes that social and behavioral interventions derived from the research agenda described below will materially improve the functioning and quality of life of older persons.

Research also emphasizes the fact that, because aging is a lifelong process, it is useful to study both the multiple determinants of successful aging and the causes and consequences of dysfunction and disability. Finally, research has demonstrated the importance of embedding the study of aging in the broader sociocultural context: race, ethnicity, gender, cultural identity, and the social environment.

The Dynamic Interaction of Individuals and Sociocultural Contexts

Genetic and other biological forces affect individual health and behavior within specific sociocultural contexts. The effects and mechanisms of these interactions affecting psychosocial aspects of health and aging need to be understood. At least four bodies of research highlight these complex dynamics. These include studies of (1) comparative aging in different societies, cultures, racial/ethnic groups, and other subpopulations; (2) the influences on aging of different environments (including geographic, workplace, and treatment environments), living arrangements, and other forms of material and social support; (3) the effects on behavioral and health outcomes of older persons' individual characteristics coupled with the varying opportunities and constraints of different social milieus; and (4) the brain/behavior relationship.

Medical research can no longer be separated artificially from sociological, economic, or psychological research. The social contexts in which individuals develop and age influence the length of their lives, their health, their ability to make decisions or cope with the demands of their conditions, their functional capacities, and the way they feel. Indeed, each of these factors influences the others. The natural laboratory provided by differences among subpopulations and subcultures in our own society, between our society and others, and between those who grew up at one time in history and those who grew up at another evidences the fact that growing up and growing old varies, for example, if one is a member of a developing society; if one is black, female, or poor; or if one is of a particular birth cohort.

Contemporary social and behavioral research in aging displays an increasingly sophisticated view of the dynamic effects of culture, social setting, and context on individual behavior (Featherman and Lerner, 1985; Maddox, and Campbell, 1985; Campbell and O'Rand, 1988; Riley, 1988; Spenner, 1988). One example of such research (Moos, 1974, 1980) illustrates efforts to measure and examine the varied settings in terms of environmental and individual characteristics and the interactions between the two. Several important conclusions have been suggested. First, the characteristics of the setting— whether structural or interpersonal factors—affect outcome. Second, personal characteristics, such as the individual capacity for cognitive appraisal and coping can, to a degree, compensate for repressive controlling environments. Third, the "fit" between personal characteristics and independence-enhancing environments predicts a beneficial health outcome.

Last, the social processes that route individuals toward beneficial milieus are as important as providing a beneficial environment in the first instance.

Differentiation

Both within and across societies, the older population is extremely heterogeneous; moreover, aging has varied substantially over historical time. People arrive at the end of life by different routes, and societies influence these routes in varied but consequential ways. Social and psychological factors affect both how individu als and the broader society age and experience aging. This variability is the best evidence for modifiability in later life. Aging is dependent on gender and ethnic, economic, political, and educational circumstances. Lifestyle, risks of illness and dependency, and functional ability also affect how one ages. Research by sociologists, economists, historians, and anthropologists has documented why references to "elderly persons," the "minority elderly persons," or "the older woman" are necessarily inaccurate and misleading (Clark and Spengler, 1980; Keith, 1985; Verbrugge, 1985; Maddox, 1987b).

Although observed variation in aging surely has a biological component, the list of explanatory factors must include those that are predominantly social and behavioral. Research has demonstrated that sanitation, food, and clean water, as well as adequate income, education, and housing, have powerful effects on life expectancy, health, and functioning. In the most impoverished developing countries, for example, average life expectancy is about half of that in developed societies (Tørrey et al., 1987). To be poor or black in the United States costs an individual, on average, 5 or 6 years of life expectancy. Poor people, who typically are also undereducated, tend to be underinsured, to use health care facilities less effectively, and to be less likely to adopt healthful lifestyles (Berkman, 1988).

In our society the differential allocation of resources—jobs or related income—by gender or ethnicity increases the risk of poverty, illness, and premature death in later life. Chronological age is only a crude predictor of behavior or health. Although the risk of functional impairment increases with age (Manton, 1988), people in their 80s have a 50 percent chance of continuing sufficiently free of disability, so as to be capable of self-care (Katz et al., 1983), although this figure decreases as one approaches the late 80s. Older individuals differ in their sense of self-control (Rodin et al., 1985; Rodin, 1986), in their capacity to cope with illness or other life challenges (Siegler and Costa, 1985), and in their risk of mental disability (Blazer and Meador, 1988; George, 1989). Such differences are dependent on social, behavioral, psychological, lifestyle, and cohort characteristics.

Modifiability

From this knowledge of differential approaches to old age stems the concept of the modifiability of aging. Improvements in socioeconomic circumstances can simultaneously improve health and well-being in the young and old. Recent research shows that many aspects of the course and outcome of the aging process are plastic—that is, open to intervention. For example, investigators have devel

oped ways in which to restore and enhance learning and memory apparently lost in very old individuals (Schaie and Willis, 1986) and to enhance self-esteem, thus enabling previously passive, institutionalized elders to take better care of themselves (Rodin et al., 1985, 1990). Because studies documenting the potential modifiability of aging have tended to focus on reducing age-related decrements in the health of individuals, research is needed to define more broadly the areas and limits of the modifiability of aging, particularly in terms of enhancing existing skills and learning in later life.

RESEARCH PRIORITIES

Three major research priorities flow from these themes.

- The first priority is investigation of the basic social and psychological processes of aging and the specific mechanisms underlying the interrelationships among social, psychological, behavioral, and biological aging functions.
- The second priority is research that addresses issues of population dynamics, including the question of whether

morbidity is being postponed commensurate with increases in longevity.

- The third priority is research that examines how social structures and changes in those structures affect aging. These priorities should be seen as the cognitive precursors to specific research questions that will examine issues related to health and functioning— the primary focus of this report. In formulating the research priorities, and in discussing the priorities listed below, the committee has not examined needed research addressing the full array of societal and specific behavioral issues and questions posed by an aging society, such as economic and political implications. The latter, however, represent an equally crucial research agenda.
- Investigation of the basic social and psychological processes of aging and the specific mechanisms underlying the interrelationships among social, psychological, behavioral, and biological aging functions should be undertaken.

A search for the mechanisms of the interrelationships among hitherto separate disciplines would include studies of (1) the social, psychological, and behavioral variables that predict health, longevity, functional ability, and well- being in individuals and (2) the most effective techniques for maintaining and improving general physical and mental health, and functioning. Once such mechanisms are identified, different intervention models need to be compared. Biologists have emphasized the inevitability and unidirectionality of decline (Rockstein and Sussman, 1979; Finch and Schneider, 1985). Psychologists, however, have found that some behavioral capacities are multidirectional: some improve over the life course, some remain stable, and others decline (Woodruff-Pak, 1988). Further, because individuals are adaptive in their behavioral adjustment to biological decline in aging, behavioral decline often is minimal. Moreover, understanding the range of limits of physiological decline and its consequence for behavior is an important priority for future research.

One of the most exciting frontiers involves the examination of the interaction among behavior, central nervous system structure and function, and neuroendocrines and other hormonal factors.

Examples include investigation of the relationship between memory function and such central nervous system structures as the hippocampus and cerebellum (Berger et al., 1986) and research on stress-related environmental, psychological, and hormonal factors that lead to anatomic changeslistb in the brain (see cerebral atrophy under extreme stress: Jensen et al., 1982; Finch, 1987). Although extensive descriptive literature exists on both physiological and psychological changes in advancing age, only limited information is available on specific behavioral correlates of specific biological changes. Data on specific interactions between biological and psychosocial factors are even more sparse. It is suspected, for example, that adverse changes in the synchrony between the autonomic and central nervous systems in advancing age may affect intellectual performance, particularly in jobs or tasks requiring complex decision making.

Despite the recognition that the major health problems of older people are chronic, little attention has been paid to behavioral and social interventions to reduce excessive disability and provide new treatment and management strategies. Some key issues for researchers include rehabilitative strategies to maintain or restore function, treatment of the social or emotional barriers to rehabilitation, and compliance with medical regimens. A variety of behavioral interventions—biofeedback in hypertension, for example—appear promising (Zarit, in press). Behavioral interventions also can be used to enhance performance of older people (Rodin, 1986).

The movement from work to retirement is one of the most important life- course decisions facing middle-aged and older workers. It has significant public policy implications, deriving from federal regulations, personnel policies, and decisions of workers. New social and economic surveys need to link business characteristics, such as pension plans and retiree health insurance, with individual career and retirement decisions. It is important to examine the age- productivity relationship in different settings (IOM, 1981; Sterns and Alexander, 1987; Spenner, 1988) so as to identify matches and mismatches between productive capacity of older workers and the personnel preferences and policies of business firms.

Adopting a life-course perspective that views today's younger adults as tomorrow's older adults should be paired with monitoring social change and the changing sociodemographic and health characteristics of populations. To the extent that children and younger adults at risk can be identified and helped, the elder population of the future will enjoy better health and well-being. Research should be designed to provide a clearer understanding of individual differences in sensory, cognitive, and behavioral aging, particularly as these differences reflect relationships between brain and behavior, environment and behavior, and society and behavior. Applied and clinical research can translate into practice what is known about the modifiability of risk factors, skills, learning, and memory. Evaluation procedures need to be improved and effective interventions devised.

- Research that addresses issues of population dynamics, including the question of whether morbidity is being postponed commensurate with increases in longevity, is a further high priority. Descriptive demography and epidemiology and techniques of population estimation have improved substantially in recent years (Hogan, 1985; Myers, 1985; Torrey et al., 1987; Guilford, 1988). Apart from the significance of immigration, two high priorities for research include (1) improved and continued forecasts of the future number of older persons, both in developed and developing nations, and (2) estimates of the proportion of those people who will be able or unable to function independently. Important work on forecasting is now under way in several research sites and is in continuing need of support. For the United States, sophisticated forecasts are now suggesting that, given a degree of control over well-known risk factors for heart disease, the size of the older population in the twenty-first century would exceed even the current high estimates (Manton, 1988). Unquestionably, the society will be confronted with unprecedented numbers of both healthy and disabled older people, and it is of critical importance that more be known about the interaction of factors— behavioral, social, and biological—that influence the time of onset of morbidity and disability as life expectancy continues to increase.

Comparative studies, such as epidemiological studies of cardiovascular disease among Japanese in different locations, are suggested as timely and useful in helping us to understand the dynamics of adaptive life transitions in varied groups. Identifying factors that enhance active life expectancy of ethnic minorities is an essential step in developing effective policy to assist minority elderly persons. Research on population dynamics needs to give special attention to at-risk populations—in particular, those subpopulations at high risk for poverty, social isolation, underemployment, inadequate education, illness, and inaccessibility to health care. Higher-priority attention should be given to such subpopulations as the poor, women, and minorities.

The physical and mental health of men and women in the United States differ in significant but paradoxical ways. Women have a higher incidence of transitory illness and a higher prevalence of nonfatal chronic diseases, they report more symptoms of illness, and they are major consumers of health care services. Women under age 65 experience more injuries, bed disability, and restricted activity days than men; this gender difference becomes even more pronounced at age 65 (Minkler and Stone, 1983). Maddox (1987a) found that women were more likely to be impaired and to be impaired earlier than men; when socioeconomic status was equalized, however, this difference disappeared. Yet, on average, women live 8 years longer than men (Verbrugge, 1985, 1988). These differences provide an obvious opportunity for cohort-sequential longitudinal research to document whether observed gender differences in morbidity and well-being are persisting or changing over time.

It is important that research also strives to identify and explain psychological variables both as predictors of health, longevity, and functional disability and as ends in themselves. Psychological and behavioral variables may not only contribute to biological aging, but may also be the behavioral representation of physiological dysfunction. In addition, certain psychological states need to be identified as conditions that must be managed in the aging process. Equally, this research necessitates detailed analyses of existing databases and the creation of new databases that contain information on health histories and psychological functioning to allow testing of the reciprocal effects of illness, lifestyles, personality variables, and psychological competence.

- Research should be undertaken to study the manner in which social structures and changes in those structures affect aging.

Just as individuals change, so too do social structures; observations of old age today show neither what old age was like in the past nor what it will be like in the future (Bengtson et al., 1985; Riley et al., 1987; Riley, 1988). It is necessary to identify stability and change in social structures and to show how stability or change affects the performance, productivity, health, and well-being of older adults. Special attention should be paid to such traditional subpopulations as ethnic and racial minorities and older women for whom social status and change entail increased risk to health. Although family structure and function continue to change, research points to a continuation of intergenerational contact and support, even when families are separated geographically (Sussman, 1985). Kinship ties still bond generations (Bengtson et al., 1985), and families continue to be the dominant source of social support in all societies (Horowitz, 1985; Hagestad, 1987).

Because of the social changes experienced in recent decades in the United States, further research is required on stability and change in living arrangements and exchange relationships. The changing structure of families and the implications of these changes for social support and caregiving demand special attention (Oppenheimer, 1982). The work force, the labor market, and the family responsibilities of women, for example, have been characterized by considerable change, but the effects of these changes have not been monitored adequately in research. The knowledge that family and labor force composition have changed does not give us the needed evidence for sound estimates of how these changes will affect the availability of social support for dependent older persons in the future. There has been little study, for example, of the implications of four-generation families, the conditions under which outside support services may affect informal family care, or how family support can be supplemented effectively by community services.

In the past decade sociologists, psychologists, and economists have looked at the relationships among work, retirement, and productivity. As our nation confronts a critical shortage of workers, increasing numbers of able older persons are spending two to three decades of their lives in retirement.

Research is needed to specify what changes in social structure could provide productive opportunities for these older people and to examine how current trends in work, retirement, and productivity are experienced by aging individuals; how they affect physical and psychosocial functioning; and how they may result in loss of productivity. Studies that contribute to the better understanding and design of assessment techniques for evaluating work capacity and prolonging work careers of older adults clearly are needed.

Additional studies are needed as well to understand the effects of the policies and decisions of business leadership on the hiring and retirement of older workers, and behavioral research is needed to discover the mechanisms that link changes in social structures (education, job experience, income patterns, and family constellations) to individual differences in maintenance of psychological competence, well-being, and adaptive capacity as people age (Schooler and Schaie, 1987; Schaie and Schooler, 1989).

Whereas the dynamics of poverty among adults are increasingly understood (Duncan, 1984), fewer data are available about the dynamics of moving into and out of poverty. Lifetime earnings, personal savings, and pension coverage provide the resources for retirement, but income and wealth can be depleted during retirement by the death of a spouse, health crises, and/or adverse financial events (Burkhauser et al., 1988). The economic status of older people has been little studied; more research on this subject is needed to untangle the complex interrelationships among aging, economics, gender, racial/ ethnic status, and health as measured by mortality, disability, chronic illness, and institutionalization.

Contextual variables are known to exert significant influence on behavior and performance. The mechanisms through which specific changes, events, or length of exposure to certain milieus influence behavioral outcome, such as role performance, health, and sense of well-being, are less clear. They must be addressed from a variety of perspectives, and caution is indicated in extrapolating from one cohort to another, particularly concerning expectations, attitudes, and preferences among older people.

How social structure may affect psychological and biological aging may be explicated as well by study of the impact of social factors upon control processes that influence an individual's sense of effectiveness over the life course (Rodin et al., 1985; Rodin et al., in press).

ADDITIONAL RESEARCH OPPORTUNITIES

These include the following:

- Social research: (1) Study the characteristics of employment, workplaces, and older individuals who are associated with continued productive activity in lifelong jobs or in new careers; (2) study how actual performance and expectations about performance of older people are affected by changing technology and examine the efforts of firms to use training and job respecification to make technological changes as age neutral as possible; and (3) develop and apply a multidimensional quality of life index to identify contributing fac tors among subpopulations for whom the index is low (e.g., the very old or minorities).
- Behavioral research: (1) Study the psychological concomitants of illness and how these affect self-care and response to formal care; (2) study the comparative effectiveness of different modalities in the treatment of chronic mental disorders; and (3) examine the effect of behavioral and social intervention on the outcome of long-term illness.

METHODOLOGICAL NEEDS

The committee suggests that three methodological approaches be used to carry out the new research agenda in behavioral and social investigation. These are as follows:

(1) refined measurement and analytic instruments such as tests, scales, and models for forecasting or analysis;

(2) cooperative multidisciplinary large-scale investigations utilizing multicultural, diverse populations studied over time and cohort sequentially; and

(3) studies of intervention techniques utilizing the findings of the foregoing research.

Tests and Instruments

Examples of new and improved measurements include (1) measures of functioning that build on current activities of living scales; (2) psychological and other behavioral assessment techniques that are age fair and culturally fair, such as tests of skills, abilities, knowledge, and capacities appropriate for older people and people in different population subgroups; (3) measures of social contexts such as home, workplace, and community; (4) more accurate ways to measure behavioral and social components of illnesses such as Alzheimer's disease; (5) improved techniques for longitudinal and cohort analysis; and (6) modeling techniques to facilitate trend prediction.

Coordination of Large-Scale Research

Although multidisciplinary research is difficult and costly, it is the only way to address many of the central issues of interrelationships and interactions among the variables and processes discussed above. These issues include demographic and epidemiological matters related to the postponement of morbidity, antecedents and consequences of health differences and sequencing among different groups, dimensions of caregiving (including effectiveness, duration, and outcome), effects of different strategies on financing long-term care, and the relation between age productivity and retirement, among others.

Field Intervention Studies

Large-scale behavioral field trials are needed to test hypotheses and evaluate recommendations for interventions based on the findings from the studies described above. These need to be systematic and of sufficient duration to observe long-term consequences and should include strategies to maintain and improve such health and cognitive functions as memory, new learning, and adaptations, as well as designs for home and workplace that would allow older people to maintain and prolong productivity and independent living. Although large-scale longitudinal research projects are necessary to implement the research agenda, the importance of small-sample studies to enhance in-depth understanding about the processes of aging must not be ignored. These studies are less costly to implement, often provide information on a relatively short-term basis, and can, for example, gather data on individual performance that are "averaged out" in larger studies.

RESOURCE RECOMMENDATIONS

Significant additional resources will be necessary to undertake the needed behavioral and social science research agenda in aging in the major priority areas. The funds to acquire these resources are described in detail in the Executive Summary and Recommendations for Funding. Despite large growth in the federal commitment to health research in the past decades, the social and behavioral sciences actually have lost ground relative to other areas of research— a trend that must be stopped if the potential of this research agenda is to be realized.

Supports for behavioral and social research on aging in 1989 were estimated at $80–$100 million from the federal government and $10–$15 million from nonfederal sources such as foundations (Behavioral and Social Research Program, NIA).

Based on the calculation of the dollars committed to behavioral and social research on aging, an estimate of the capacity for high-quality research, and the additional monies necessary to develop the new initiatives identified here, the total budget for age-related behavioral and social research should be increased by more than 100 percent over current expenditures, with the additional funds to be phased in over a 5-year period.

The recommended distribution of additional funds for the research initiatives should support studies of social, psychological, behavioral, and biological interrelationships; "at-risk" populations; population dynamics; and changing social structures.

These resources involve (1) increasing the funding rate of NIA and other NIH approved research program grants (RPGs) on aging from one in four (24 percent) to one in two; (2) providing training for an additional 200 investigators per year in behavioral and social science; and (3) a one-time cost for construction. New resources will include a share in ten multidisciplinary research and training centers added to the current three centers (Claude Pepper Centers) supported by the NIA and participation in an expanded scientific infrastructure (laboratories, computer support, long-range population studies, and increased linkage to such existing databases as the Health Care Financing Administration Medicare data tapes).

The increasing isolation of older persons (one-third live alone), the burden of psychologically disabled older persons on families, and the presence of serious affective disorders (older white males have the highest suicide rate) among elderly persons point up the urgent need for research that leads to better management of social and behavioral problems in the older population.

CROSS-DISCIPLINARY AND CROSSCUTTING ISSUES

Basic biomedical and clinical researchers have only recently begun to examine the influence of gender, race, culture, ethnicity, and social class on biological processes and clinical presentations. These and other social and behavioral issues cut across the disciplines and offer opportunities for further investigation and understanding in the field of aging.

Gender

The difference in life expectancy at birth between the sexes was small for those born in 1900, but it has widened considerably over the century. Much of this difference can be ascribed to changes in child-bearing practice and other lifestyle factors as well as to tobacco-related deaths and heart disease.

Social and behavioral gender differences in morbidity, particularly as related to chronic conditions, suggest many promising research areas, such as the gender- specific response to social isolation and psychological disorders; differences in the relation of social supports to functional status; the differential effects of gender in pension income, social role expectations, labor force participation, and responsibilities for caregiving; interactions among neuroendocrine and other hormones; and cognitive and psychological status. Potentially significant are the many connections of this research with work in clinical science and health services delivery research as it relates to alterability of risk factors, appropriateness and effectiveness of interventions, utilization, service delivery, and many other areas.

Race and Ethnicity, Socioeconomic Status, and Other Cultural Factors

As is the case with gender, the cultural factors associated with race and ethnicity present important variables across and among the disciplines of aging. For example, racial and ethnic differences in metabolizing certain drugs and substances are confounded by cultural variables in the use of these substances, definition and presentation of symptoms, other illness behaviors, and acceptance of treatment regimens.

In addition, an understanding of the effects of racial and ethnic-cultural differences in resource allocation is critical in explaining clinical, social, and psychological differences in aging and in health care service utilization. Moreover, race and ethnicity may well be contributing factors to underlying basic biological processes or their manifestations in old age.

Population Dynamics

Issues of population dynamics include accurate estimation and prediction of population trends in longevity, mortality, morbidity, and functional status. These dynamics also have major linkages with health services delivery concerns about access, availability of services, financing, system design, and policy formation. However, these issues, particularly as they concern estimates of the health and functional problems most likely to face large numbers of elderly people in the coming decades, also have connections to basic biomedical and clinical research in terms of predicting problems and defining important areas of research.

Brain, Environment, Society, and Behavior

Some of the age-related deficits in memory performance and cognitive function are amenable to understanding at the neurobiological level. But the question of the extent of memory deficit due to deterioration of basic storage mechanisms in the brain in contrast to those due to altered modulatory processes also requires a better understanding of the related psychological and psychosocial processes.

Moreover, the interaction of the brain with aspects of the environment—physiological, physical, and social—is also poorly understood (Finch, 1987). For example, environmental stress can influence sleep patterns, which can then reflect on brain activity and the mental and emotional sequelae. Such interactions demand both crosscutting and cross-disciplinary approaches to further our essential knowledge and understanding.

4

Health Services Delivery

The existing health care system is fraught with problems for older persons and is best described as chaotic—disorganized, inadequate, and poorly distributed. There are major gaps separating acute and chronic care of older patients, and there is no organized array of services to meet these patients' chronic care needs or to provide for continuity of care. Equally, the health care system lacks the sensitivity to respond to the changing needs of individual older persons as they experience transitions in health. Although this report focuses on the health needs of the geriatric population, it must not be forgotten that many of the defects in long-term and chronic care apply to younger adults as well. In the 1990s health care services will continue to be challenged by the changing demands imposed by the projected growth of the older population and especially by the growth of the "oldest old"—those persons 85 years of age and older who are most vulnerable to the problems of ill health and dependency.

The growth of the older population will accelerate even more at the beginning of the next century as the baby-boom generation ages. Older persons consume an enormous amount of health services. Over the next decade those persons 80 years of age or older will likely become the largest single federal entitlement group, consuming $82.8 billion annually, including social security and other benefits (Torrey, 1985). Unfortunately, the response to this growing challenge has been far from adequate.

For example, the rate of production of health professionals (including those in health services delivery) is far short of the goals to provide care for dependent and chronically ill older persons and to meet future needs (National Institute on Aging, 1987). This deficiency will be compounded further by the loss of informal supports as more and more women join the labor force, therefore making them less available to provide care for older relatives. Three trends are likely to influence health policy and health services for older persons in the 1990s: (1) the continuing problem of rising health care costs, (2) the growing concern with quality of care and cost-benefit ratios, and (3) the continuing rapid structural change of the health care service delivery system (Gilford, 1988).

Health services research focuses on broadening the knowledge base required to understand and influence the organization, delivery, and financing of health services to older persons. This research area cuts across the other four agenda areas in this report, relying heavily on knowledge from basic research in the biological and psychosocial arenas and linking with clinical research in studies designed to test delivery strategies for proven interventions in a variety of settings.

By studying and making more effective the factors that influence delivery of health care to older individuals, research in health services delivery can contribute to decreasing disability and chronic illness, and to improving the quality of life of the older generation.

The following issues in health services delivery were identified by the committee as having greatest relevance to the research priorities listed in the next section:

- long-term care and continuity of care for older persons;
- financing health care for older persons;
- drug use and mental health services for older persons; and
- disability/disease prevention and health promotion services for older persons.

Implicit in research involving these areas are questions regarding (1) the extent to which functional status and/or quality of life for older persons is improved with various treatment options and (2) the efficacy and cost of care. This focus on effectiveness, outcome, and functional independence fits well with the new direction being taken by federal research centers.

An extended discussion of the issues related to health services research in aging appears in the Health Services Delivery Research Liaison Team report (see Appendix B).

LONG-TERM CARE AND CONTINUITY OF CARE FOR OLDER PERSONS

While there remain many unanswered questions concerning health services research related to older persons, the most pressing are those involving continuity of care, particularly with respect to long-term care. The objectives of long-term care are to help older persons cope with their disabilities, decrease dependence on others, and narrow the gap between actual and potential functioning (U.S. Department of Health and Human Services, 1980). Our system for delivering and financing long-term care is in an embryonic developmental state. A majority of the long-term care needs of this population are unmet or met by families (particularly the female spouse or daughters of the disabled older person), clearly indicating the need for governmental support of these important caregivers.

Long-term care is defined in a number of ways. The Health Care Financing Administration, the major federal agency involved in payment for long-term care for older persons, defines it as follows: "Long-term care refers to health, social and residential services provided to chronically disabled persons over an extended period of time. The need for long-term care is not necessarily identified with a particular diagnosis, but rather physical or mental disabilities that impair function in activities of daily living" (Doty et al., 1985).

A second definition can be found in the introduction to the Long-term Health Care Minimum Data Set: "Long-term health care refers to the professional or personal services required on a recurring or continuous basis by an individual because of chronic or permanent physical or mental impairment. These services may be provided in a variety of settings including the client's home" (U.S. Department of Health and Human Services, 1980). These and other definitions of long-term care (Kane and Kane, 1987) share two common elements: (1) the care continues over an extended period of time and provides varying degrees of intensity and resource deployment; (2) it is provided to persons who have lost (or never had) the capacity to care for themselves because of a chronic physical or mental illness or condition (Gilford, 1988). The transition between acute and long-term care is complicated because the mechanisms that provide continuity of care for patients moving between acute and long-term care are not well developed. Consequently, issues surrounding continuity of care continue to plague older persons, their families, health care providers, and researchers.

Characteristics of Older Persons Requiring Long-Term Care

As noted in the Executive Summary and Recommendations for Funding, chronic diseases have become major contributors to death and disability among older persons in this country. However, the Panel for Statistics for an Aging Population (Gilford, 1988) states that "the diagnostic description of a person is seldom an indicator of his or her need for or utilization of long-term care because diagnosis alone gives no clue to how well or how poorly an individual functions. As a result, descriptions of behavior, rather than conditions, are used to describe the long-term care population."

About 20 percent of those 65 years of age and older need assistance to carry out some basic functions of adult life (LaPlante, 1988; Rice and LaPlante, 1988). Because of the dramatic and varied effect of age on disability, morbidity, and mortality, and because the older population is a widely heterogeneous group, investigators have suggested dividing this population into age subgroups. Gilford (1988), for instance, recommends three subgroups for health research purposes: the young-old (65 to 74 years), the old (75 to 84 years), and the oldest- old (85 years of age and over). This last group is most vulnerable to disability and the problems associated with ill health and frailty, and is most in need of long-term care services.

Data collection related to the natural history of chronic illness and disability, functional status, behavioral problems, and quality of life should be supported and expanded, with particular attention paid to changing cohorts (one- third of the membership in the aged population changes every 5 years). The recent mandate to collect functional status data on nursing home residents will provide a needed database on that population. Much more research is needed to elucidate the characteristics of special populations of older persons requiring long-term care. For example, little is known about how ethnic and class differences influence the need for or access to care.

Informal Providers of Long-Term Care

Family members and friends provide over 80 percent of all long-term care for older persons. Most caregivers are women; about 60 percent are 65 years old or older. More than 40 percent of those giving care rate their health as fair or poor, and over a third rate their health as poor or near poor (Stone et al., 1987). The past decade of research indicates that family caregiving is a complex phenomenon. Tremendous variability exists in the nature of the care provided by family members, the stress and rewards related to that care on the part of the caregiver and older person, and the relationship between the formal and informal systems of care. Caregiving can be stressful for the caregiver and other family members (Zarit et al., 1986). In fact, recent research has shown that family members of older persons suffering from mental health problems are at risk for developing their own mental disorders (National Institute of Mental Health, 1988).

Those family members providing care to older family members receive little help from the formal health care system, for policymakers are concerned that extending publicly supported services to caregiving families might reduce the amount of family care provided to older persons. However, available evidence in this area suggests the opposite: the addition of formal services to a family caregiving situation does *not* reduce the amount of informal help provided to the older person (Stephens and Christianson, 1986).

Although our understanding of the phenomenon of family caregiving is increasing, we lack knowledge regarding caregiving among special populations (e.g., minority elderly persons, low-income elderly persons, elderly persons living alone), appropriate interventions for caregivers, and targeting interventions for this group (Brody, 1990). The nation depends on the family to provide most long-term care services in the community, but the knowledge base regarding how to assist family caregivers is inadequate.

Formal Long-Term Care

Organized long-term care in this country has been criticized as expensive, inadequate, disorganized, and biased toward institutional care (Harrington et al., 1985; Kane and Kane, 1987). The continuity of care between acute and long- term care is not well developed and often nonexistent (Estes, 1988). During the past decade, attempts have been made to address these problems in long-term care. Results of these long-term care demonstrations (e.g., the channeling experiments) are difficult to interpret because of methodological problems encountered in the studies, but the hope that high- quality long-term care could be paid for with dollars recouped from the prevention of hospital care was not realized (Weissert et al., 1988). Several criticisms have been directed at past long-term care experiments: (1) they failed to account for the complex and changing nature of long-term care needs; (2) the outcome measures used to evaluate such interventions were inappropriate; and (3) criteria for the selection of persons receiving such interventions were not appropriate. Results of long-term care demonstrations, however, do sug gest an avenue for further research: both the conceptualization of long-term care and the methods used to study the efficacy of such services need to be reviewed. In addition, management mechanisms, such as case management, that are designed to facilitate continuity of care need to be evaluated.

Supply of Personnel for Long-Term Care

As our population ages, institutional and community-based long-term care systems are and will continue to be confronted with increasing numbers of older persons. Without intervention the recruitment, training, motivation, and reward of a competent work force to meet this population's service needs will continue to be insufficient. To a great degree this problem represents an economic and personnel issue, not a research question. The central questions are the following: Who should provide long-term care, and how can we pay the providers? To the extent that research can elucidate underlying factors affecting the choice of long-term care careers, work satisfaction, job design, and other work force-related matters, it should be supported.

Since long-term care providers are generally women, many are also poor and many are members of minority groups. From the provider's perspective the issues confronting workers in long-term care settings are "women's issues "— poor salaries and benefits, low job security, and lack of flexibility in scheduling. In many states, in fact, nonprofessional providers of long-term care are paid no better than the minimum wage.

Additionally, the gerontological and long-term care content of traditional physician, nurse, and social worker educational programs is inadequate (National Institute on Aging, 1987). Numerous professionals are involved in delivering health care to older persons; their roles, however, are not clearly defined. Research focusing on role definition and the costs and benefits of alternative long-term care providers should be undertaken. In order to avert a crisis within the next decade, research and demonstration programs that focus on the recruitment, training, retention, and quality of both nonprofessionals and professionals in long-term care are needed immediately.

Locus of Long-Term Care

At this point, little agreement exists about the most appropriate locus of long-term care. Is it the hospital, the home, the nursing home, or another long- term care facility? Research is needed to evaluate the attributes of each of these settings in relation to long-term care, the attributes of persons providing care within these settings, and the ability of specific settings to adjust to the changing needs and personal resources of older persons.

Quality Assurance

The interactions among the need for care (based on functional status), the nature of long-term care intervention, and the outcome of care are poorly understood. Lacking are standards for long-term care and valid and reliable measures of the need for and quality of care, including quality of care at different sites (e.g., home care), and instances where quality of care may conflict with quality of life. Development of such standards and measures for community-based care is particularly critical because of the shift in locus of care to the community. Such measures include indices of the quality of family care along with the quality of formal community care.

Several issues that cross health services and clinical research areas require attention. First, information about how a provider's perception of age and aging might affect treatment decisions in long-term care is essential. Second, research focusing on understanding rehabilitation strategies and their relationship to treatment outcomes is needed.

Technologies

Recent technical advances have the potential to enhance life for many disabled people, but little work has been done to delineate the positive and negative effects of such technologies. Three categories of technology are relevant to long-term care: (1) universal technologies (those shaping the environment in home and community to enhance, or to inhibit, independence), (2) individual technologies (those providing aids to individuals to help overcome deficits), and (3) life-sustaining technologies (those replacing or supplementing failed organs or bodily systems). The use of these technologies is value laden and leads to central questions that have not been addressed in our society. This and other ethical issues are discussed further in Chapter 6. Theoretical development in the area of universal technologies suggests that such strategies may be key to reducing the need for both individual aids and service interventions (Orleans and Orleans, 1985; Zola, 1988). This area needs more research attention and represents a promising research direction.

FINANCING OF HEALTH CARE FOR OLDER PERSONS

Because of increased health care expenditures by federal and state governments and by individual older persons, health care financing has become

one of the most critical issues to be addressed by the nation. Spending on health care services increased at an annual rate of 14.5 percent between 1977 and 1984 (Waldo and Lazenby, 1984) and has continued to increase since that time, but there is no evidence of a comparative improvement in health outcome. To a great degree, the methods used to finance health services govern both the nature of the services used by and available to older persons and the growth and development of the health and social service systems per se (Gilford, 1988). Policymakers and health services researchers are confronted by five major issues related to the cost and financing of health care for older persons:

(1) the continuing increase in the cost of health care despite efforts to restrain spending;

(2) the highly concentrated distribution of health care expenses in which very few persons actually consume the majority of health care resources;

(3) the individual and societal consequences of a health care policy that does not cover the costs of long-term care;

(4) the impact of poverty on health status and access to care; and

(5) the lack of suitable health care outcome measures.

These issues are complicated further by the fact that positive health is not distributed equally across society. Poverty negatively affects health; ill health impoverishes. Consequently, older persons most in need of health care are vulnerable not only because of ill health but also because of poverty. Since poverty is not distributed equally throughout the population but is more prevalent among women and members of minority groups, older persons who are poor, female, and members of a minority group are most vulnerable to the problems of illness.

These matters require research to help resolve the problems of health care financing for older persons. First, the factors that determine patterns of public and private health insurance arrangements for older persons must be understood of particular interest is the problem encountered by many older persons who have multiple insurance policies, overlapping or inadequate coverage, or health policies as part of the retirement benefits provided by government and industry.

The failure of this fractured structure to protect many of the aged who are chronically ill stresses the urgency for intensive research on long-term care insurance for this vulnerable population.

Second, the effects of financial barriers on access to and use of health services and on the health status of older persons must be better understood. Third, research related to the distribution of risk should be approached from a theoretical framework to guide definition and selection of risk factors and data collection. At present, models are tested using data that may not be appropriate. For example, investigators limited by inadequate research support have been forced to use data collected earlier by Medicare; these data often fail to identify the presence or absence of family caregivers or other supports for older persons. More work is needed to examine how financing and regulations shape the care system (by providing incentives for sites and types of care) and how Medicare policies influence screening and preventive services for older clients. Research should be directed toward better understanding providers who demonstrate inappropriate responses to older persons. Finally, more study is needed of careers in chronic care and of the cost of different levels and patterns of chronic care.

DRUG USE

Whereas persons 65 years old and over comprise about 12 percent of the U.S. population, they receive about 25 percent of the total number of prescriptions dispensed, and they account for about 30 percent of annual U.S. prescription drug expenditures. Three major and well-founded concerns arise in the area of geriatric drug use: inappropriate physician prescribing, widespread elderly patient noncompliance with drug regimens, and increasing drug prices.

Inappropriate Prescribing

Our body of knowledge regarding the effects of aging on drug therapy and drug metabolism is extensive (Bender, 1974; Lamy, 1984; Lipton and Lee, 1988; Montamat et al., 1989). However, increasing evidence indicates that prescribers do not follow the guidelines set forth in this research. Thus, a number of problems arise in the process of prescribing: a drug may be ordered when no drug is needed, or a prescription may be made for the wrong drug, a suboptimal drug, the incorrect dosage, or a dosage without knowledge of other medications being taken.

Evidence of geriatric prescribing problems can be found in the high rate of drug-related hospitalizations among older people. Among all patients admitted to the hospital, patients 65 years old and older have the highest admission rates (10 to 25 percent) for drug-induced symptoms, and this problem appears not to have changed significantly during the past decade (Levy et al., 1979; Bergman and Wilholm, 1981; Ives et al., 1987). Recent studies focusing on geriatric patients suggest that 10 to 31 percent of hospital admissions are associated with drug-related problems (Williamson and Chopin, 1980; Popplewell and Henschke, 1982; Grymonpre et al., 1988). Drug-induced hospitalizations are associated with appreciable morbidity and mortality; the cost of such hospital care is estimated at $4.5 to $7 billion annually (Kusserow, 1989).

Although most of the studies identify specific drug classes implicated in drug admissions, they do not identify the specific reasons the drugs are associated with the problem. For example, if a given drug is responsible for a hospital admission, is the problem related to drug dosage, drug schedule, drug allergy, drug duplication, drug-drug interaction, or drug selection? Additional research is needed to add to our understanding of the mechanisms through which specific drugs lead to problems associated with hospitalization. We also need to learn more about the incidence of prescribing problems among older persons living in the community. The cumulative burden of drug-related morbidity among elderly outpatients may have enormous clinical and economic consequences.

Noncompliance

Older persons are more likely than younger persons to misuse medications (Cooper et al., 1978; Lamy, 1984; Darnell et al., 1986). A systematic investigation of communication among patients and providers should be undertaken to assess the extent to which good communication enhances compliance on the part of older persons (Lipton and Lee, 1988). Systems of information retrieval for physicians, pharmacists, and other providers should be developed and evaluated. Finally, demonstrations are needed to test approaches to improve compliance by older populations. Because the aging process alters pharmacodynamics and pharmacokinetics, older persons are at risk for untoward drug effects (Law and Chalmers, 1976; Kiernan and Isaacs, 1981). In addition, many older persons with multiple chronic illnesses take more than one drug, increasing the chance of an adverse drug reaction (Steel et al., 1981; Hutchinson et al., 1986). While we know that older persons are at risk for drug reactions for several reasons, we do not know the extent to which older persons are at risk. Work by Lipton and Lee (1988) identifies multiple strategies to reduce these risks through such broad-based interventions as modified labeling and simplified drug schedules.

Drug Costs and Older Persons

National expenditures for "drugs and medical sundries"—a category that includes prescription and nonprescription drugs and medical supplies dispensed through retail channels—accounted for almost 7 percent of health spending or $34 billion in 1987. Drug costs totaled about $9 billion for older persons that year.

The number of drugs prescribed increases with age, and the average price of prescriptions also rises with age. Overall, older persons pay 14 percent more per prescription than do those under age 65. It is important to note that older persons are not being charged more for their medications. Rather, they take a different mix of drugs, and their prescriptions extend over a long-term, not acute, treatment period. Because they use more drugs and pay higher prices for prescriptions than do younger patients, older patients are more likely than other age groups to incur high drug costs. They are also more likely to pay a higher percentage of out-of-pocket expenses for drugs.

What about the demographic distribution of drug expenses incurred by older persons? Results of a 1980 national survey revealed that the proportion of prescription drug expenses reimbursed by private insurance was similar for both poor and near poor older persons: 7.7 percent and 10.7 percent, respectively. However, that statistic rises significantly—up to 20.6 percent—for those who are not poor. All three groups had out-of-pocket expenses of approximately half of the total costs of drugs. The near poor (those living in families whose income is above the poverty level but less than or equal to twice the poverty level) are particularly vulnerable to this out-of-pocket cost burden.

It is important to emphasize that this information was collected in 1980, before the escalation of drug prices. Since 1981, prescription drug prices have increased two to three times faster than all other consumer prices (Lipton and Lee, 1988). As a result, drug prices have become a source of growing concern for consumers, policymakers, and both individual and institutional purchasers. Although the real income of the aged poor and near poor has remained relatively static in recent years, with Social Security payments increasing at the same rate as inflation, the expenditure for drugs most likely has increased as the result of drug price inflation. Therefore, drug expenditures relative to income may be even greater now than at the time of the 1980 survey.

MENTAL HEALTH SERVICES

There is growing interest in the older person's use of mental health services, with some researchers noting that older persons seem to use fewer mental health services than do younger persons (German et al., 1985; Leaf et al., 1985; Borson et al., 1986; German et al., 1987; Goldstrom et al., 1987; Lurie and Swan, 1987). The need to understand the older person's use (or nonuse) of mental health services is important for several reasons: (1) the high incidence of dementia in the older population, (2) the high prevalence of mental impairment of older persons in nursing homes, and (3) the high prevalence of alcohol abuse or dependency among older persons admitted to county or state mental hospitals. In the community 10 to 25 percent of older persons have some degree of mental impairment. Among men, the incidence of suicide dramatically increases with age. In short, mental illness is common in older persons, but often is untreated. Research is needed to investigate how older persons use mental health services and why of particular interest are the effects of race, gender, and social status on service utilization.

Alzheimer's disease is widely recognized as the major mental health problem of the older population, but little is known about its epidemiology and treatment of particular interest is the delineation of risk factors associated with dementia. Depression is another common problem of late adulthood; yet because there is little agreement on the definition of this condition, depression in older people may not be recognized widely. From a health services perspective, innovative methods for the treatment of depression that are acceptable to older persons should be developed and evaluated.

Our understanding of the less common clinical problems of aging— schizophrenia, bipolar disease, and other psychotic conditions—is very limited. Research is needed to examine both those who develop a psychotic condition, such as paranoia, in older age and those who were diagnosed with such disorders earlier in life. Even where our knowledge base is firm, clinicians providing the bulk of services to older persons may not recognize the signs and symptoms of major mental illness.

Investigation is critically needed on mental illness in special populations of older persons. Little is known, for example, about the mental health status or needs of older persons who are members of minority groups of the older population, those who are institutionalized have the least understood but highest prevalence of mental disorders (National Institute of Mental Health, 1988). We do not know what services are available to this group, nor do we know what services are needed or how they should be financed. Research also is needed to elucidate the interactive processes among disability, disease, mental health, the environment, and the cost of health services. Such a holistic approach should lead to a better understanding of the panoply of services needed by older persons. Finally, research is required to elucidate the cost and outcomes of various patterns of mental health care. How well, for example, do community mental health centers meet the needs of older persons? These data should be related to the reimbursement system that currently does not pay for many types of mental health services.

DISABILITY/DISEASE PREVENTION AND HEALTH PROMOTION SERVICES

Although some work has been undertaken in the area of disease prevention through risk reduction,* there is still a crucial need for clinical research in this area. Lacking is consistent evidence regarding the effects of risk reduction on the presence of disease conditions; no systematic research has been undertaken on risk reduction in late adulthood. Lifestyle prescriptions for older populations are now based on research done on younger populations.

Several clinical research questions need to be answered before health services research can proceed:

- Are accepted risk factors for disease the same in direction and magnitude for older persons as they are for younger persons?
- What are the prevalence estimates for specific risk factors by demographics within the older population?
- Are interventions known to be effective with younger persons equally effective and acceptable in reducing risk in older people?
- Does the reduction of risk factors make any difference in the health status of older persons?
- What is the relationship between disease and disability?

If research does support the need for intervention, then demonstrations designed to deliver preventive services to older persons should be developed. Research also is needed to identify risk factors for disability; risk-reduction programs can then be developed and evaluated.

* See the Institute of Medicine's 1990 report, *The Second Fifty Years: Promoting Health and Preventing Disability* (Washington, D.C.: National Academy Press).

RESEARCH PRIORITIES

The committee identified five research priorities in the area of health services research. These priorities represent the broad areas within which specific research questions and proposals can be framed. It is noteworthy that the field of health services delivery, as with some aspects of clinical science and behavioral and social studies, offers opportunities for research in nursing and other aspects of health care that may contribute significantly to the field of investigation of aging.

- **The committee proposes that research be undertaken into long-term** care and continuity of care for older persons.

  This priority area includes research into factors that determine the need for and use of long-term care services, questions regarding the organization and delivery of long-term care services, the special role of the family in the delivery of care, and the growing question of adapting modern technology to enhance the ability of older persons to live independently. Also included in this category is research into the nature of the hospital as an institution and the role of nursing facilities and health care institutions for the older person. Finally, this priority area includes research at three levels—the individual, the institution, and the society—that affect recruitment, training, and quality of providers of both formal and informal long-term care.

- **The committee recommends research on the cost and financing of** health care in the older population.

  This priority area includes research into reducing financial barriers to care for older persons of particular interest are questions related to financing of insurance for older persons, most notably coverage for long-term care, mental health, rehabilitation, and disease prevention and health promotion services. Studies of outcome measures in this area also are important.

- **The committee recommends research on drug therapy in older** persons.

  This priority area includes research into the factors influencing the development and regulatory approval of drugs and the effectiveness and efficiency of prescribing, dispensing, and taking pharmaceuticals. Equally, research should be undertaken into drug-related phenomena that affect the social and physical functioning of older persons.

- **The committee recommends research on mental health services in** the older population.

  This priority area includes research into the factors that determine the need for and use of mental health services, the role of treatment for mental illness as a factor in improving the social and physical functioning of older persons, and the effect of early life treatment in reducing later-life disability from mental disorders.

- **The committee recommends research on disability/ disease** prevention and health promotion in older persons.

  This priority area includes research into developing new models for the delivery of appropriate behavior change programs for elderly populations.

ADDITIONAL RESEARCH OPPORTUNITIES

The committee considers the preceding five research priorities to be the most critical. Additional important opportunities in health services delivery research include research into alternative methods of delivering long-term care, including models that focus on a particular subgroup (e.g., posthospital management) and that provide linkage of housing arrangements with community long-term care facilities; research to examine the effect of employer-initiated cost sharing on the use of preventive and treatment services by older persons; research on the effect of interventions to reduce the incidence of inappropriate drug treatment decisions by physicians; research to study the provision of mental health services to elderly institutionalized persons in the context of need, utility, and outcome; and research to investigate and compare the relative contribution of accepted risk factors in the development of disease among young and aged populations.

RESOURCES REQUIRED

The committee recommends increased support over current health services research spending during the next 5 years for increased funding of research, for construction, for training, for the expansion of infrastructure (databases, library support, computer capability), and for the addition of new Centers of Excellence in Geriatric and Gerontological Research and Training (Claude Pepper Centers). Funding needs are described in greater detail in the Executive Summary and Recommendations for Funding. Specifically, increased funding in age-related research is recommended in order to raise the number of approved grants that are funded from less than one in four (24 percent) to one in two. In addition to the dollars for research activities, funds are needed to support a minimum of 140 additional predoctoral and postdoctoral fellowships per year for training in health services research careers, for specific infrastructure needs, and for construction costs. Health services research also will participate in the added funds for support of programs of geriatric and gerontological research that are located within current medical centers (administrative locus, as compared with geographically established centers).

CROSS-DISCIPLINARY AND CROSSCUTTING ISSUES

Most of the research priorities identified in this chapter cross disciplinary lines, focusing on issues that benefit from clinical, behavioral, and social science as well as health services research. Beyond that, four key crosscutting issues must be addressed directly.

First, an understanding of the effects of gender, class, and ethnicity is salient to every priority area listed in this chapter. Second, ethical issues that must be thoughtfully analyzed and researched are embedded in each priority. Third, save in specific areas, health services research in the area of health promotion/ disease prevention cannot proceed until clinical research provides clear mechanisms for risk reduction in the older age groups. This need has been identified by Lipton and Lee (1988) who state with respect to drug use research: Despite the existence of a great deal of information regarding drug therapy and older patients (especially information gathered in recent years), there is a need for more research. Areas requiring further study include the nature of age- related biological, physiological and pathological changes; ways in which these changes affect the elderly person's response to drugs; and the kinds of drug prescribing, dispensing, and administration appropriate to deal with these changes. An area that has remained virtually unexamined involves the psychological changes (e.g., depression) and social changes (e.g., loss of spouse) that accompany aging and the way they affect the older person's need for, use of, and response to drugs. Drug epidemiology studies—large-scale studies of drug use and its relationship to clinical outcomes—are urgently needed in elderly populations, especially with regard to psychotropic agents. Private foundations and federal funding agencies (particularly the National Institutes of Health and the Alcohol, Drug Abuse and Mental Health Administration) should give these areas high priority on their agendas.

Fourth, the dynamic nature of both the aging process and the health care delivery system means that there is a great need for longitudinal research. Most of the research areas in this section can best be addressed using this approach. Longitudinal research is complicated by a series of predictable methodological and situational issues that must be taken into account when planning for resources in this area.

5

The Biomedical Ethics

Individual older persons and the aging American society as a whole face dilemmas regarding life-sustaining treatment, distribution of limited health care resources, and participation in clinical research by vulnerable elderly persons. Research can help clarify the underlying ethical issues, analyze the arguments for and against various stances, and forge a social consensus to guide clinical care and overall health care policy. Furthermore, empirical studies in biomedical ethics can help identify in current practice any variance from recommended guidelines, and develop guidelines and evaluate interventions to improve clinical practice.

Although the topics and research recommendations developed in this chapter focus on the problems of older persons, many of these issues (e.g., decreased mental competence, need for advance directives, disabled patients without families or friends) concern younger disabled persons as well.

RESEARCH PRIORITIES

Sufficient funds should be made available to conduct research on the ethical dilemmas involved in the provision of life-sustaining treatment, the allocation of health care resources, and the participation in clinical research by frail and elderly persons. Using the broad areas of investigation listed here, one can formulate specific research questions to include both empirical and analytical studies.

Life-Sustaining Treatment in Older Persons

- **The committee recommends research on decisions regarding life-** sustaining treatment in older persons.

The medical technology that allows physicians to prolong life in patients with serious illnesses may not always be appropriate. According to ethical, medical, and legal guidelines, competent informed patients may refuse life- sustaining treatments. For patients who lack the capacity to give informed consent or refusal, such as elderly patients with severe dementia, decisions concerning medical care should respect wishes previously expressed by the patient (President's Commission for the Study of Ethical Problems in Medicine and Biomedical and Behavioral Research, 1982, 1983; Annas and Glantz, 1986a; American College of Physicians, 1989a, b). Such declarations by competent patients of what care they would or would not want if they became incompetent are called "advance directives."

However, these guidelines may be problematic for elderly persons. Although most elderly persons welcome the opportunity to discuss their preferences for (or against) life-sustaining treatment, they infrequently develop or invoke specific advance directives (Lo et al., 1986).

Further knowledge is needed about ethical questions concerning both those isolated elderly patients who are institutionalized and those older persons who are not institutionalized, but who do not have relatives or close friends to act as surrogates. (It is important to note that the number of isolated older people most likely will continue to grow in the coming decades.)

Improving Discussions About Life-Sustaining Treatment

Research is required to lay the foundation for increased understanding of what requires improvement in decisions about life-sustaining treatment among physicians, families, and institutions. Educational interventions that encourage health care providers and patients to discuss life-sustaining treatments need to be developed. In the absence of explicit discussions, physicians and family members cannot predict a patient's preferences for (or against) treatment (Bedell and Delbanco, 1984; Uhlmann et al., 1988). Such discussions might be more effective if more were known about how physicians currently engage in them. More needs to be known about how well elderly patients understand the issues at stake, how their understanding might be improved, how their preferences might change over time, and how they can be given specific directives about life-sustaining treatments.

The focus of research on older patients' acceptance or rejection of life- sustaining treatment may vary, depending on whether the decision involves patients and their families or is limited to isolated institutionalized and noninstitutionalized older persons.

Several problems with advance directives must be resolved. The durable power of attorney for health care is the most effective type of advance directive because it allows competent patients both to designate surrogates and to specify the types of treatments they do and do not want (President's Commission for the Study of Ethical Problems in Medicine and Biomedical and Behavioral Research, 1983; Buchanan and Brock, 1986; Force, 1988). However, such advance directives cannot be used by patients with no relatives or friends to be designated as surrogate decision makers. Living wills are ineffective for such patients because they apply only in cases of terminal illness and are inapplicable to such situations as severe dementia (President's Commission for the Study of Ethical Problems in Medicine and Biomedical and Behavioral Research, 1983; Buchanan and Brock, 1986; Force, 1988). Thus, to ensure that patients' preferences are respected if they become incompetent, new procedures need to be devised for patients with no one to serve as a surrogate. Because legal problems often arise in this context, as well as in other areas reviewed in this chapter, these problems present an opportunity for collaborative research between medical and legal scholars.

Improving Decision Making for Incompetent Patients Who Have Not Given Clear Advance Directives

Such situations are common and perplexing. Consensus has not been reached regarding either the level of specificity required by an advance directive or the discretion afforded family members in interpreting such directives (Veatch, 1984; Rhoden, 1988). Further, when the incompetent patient's preferences are unclear or unknown, no specific guidelines delineate how the patient's best interests should be determined, what role quality of life considerations should play, and what safeguards are desirable for particularly vulnerable nursing home residents (Buchanan and Brock, 1986; Lo and Dornbrand, 1986; Lo, 1988). In addition, research is indicated on how advance directives play out—behaviorally and legally—to inform revision in ethical standards and law. Further analysis of these issues is needed; discrepancies among clinical practice, ethical guidelines, and case law must be studied.

Defining the Role of Institutional Ethics Committees, Particularly Within Nursing Homes

Such committees have been proposed to improve decision making about life-sustaining treatments and to serve as an alternative to time-consuming and cumbersome legal proceedings (Cranford and Doudera, 1984). However, more must be learned about the benefits and drawbacks of such committees, their optimal structure and procedures, and their impact on clinical decisions (Lo, 1987).

Establishing Medical Futility

Physicians and other providers sometimes decide to withhold treatment because it is considered medically futile. Better data are needed on prognosis and outcome in seriously ill elderly patients. Although some good data are available about cardiopulmonary resuscitation (Bedell and Delbanco, 1984), many other life-sustaining treatments should be studied, particularly those affecting the critically ill and patients of very advanced age (over age 85) (Force, 1988). In addition, the concept of futility itself should be analyzed. The term can be used in many ways, some of which do not constitute acceptable justification for physicians to withhold treatment (Youngner, 1988).

Resolving Disagreements Between Health Care Professionals and Patients (or Patients' Families)

Caregivers may insist on giving a treatment—particularly artificial feedings —despite the refusal of the patient or family (Steinbrook and Lo, 1988; Miles et al., 1989). Court rulings on cases involving such disagreements are often misunderstood by physicians, such as interpreting the Supreme Court's ruling on nasogastric feedings in the Cruzan case to mean that nasogastric feedings can *never* be discontinued in incompetent patients (Kapp and Lo, 1986; Annas, 1990; Lo et al., 1990). Formal study should be undertaken to learn more about how physician perception of legal requirements and liability affects decision making, and further legal and ethical analyses are needed to begin to develop a consensus on how best to resolve these dilemmas. An area of research involving disagreements between caregivers and patients that deserves further attention, although it relates more to patient comfort and contentment than to life-and- death issues, involves study of day-to-day ethical problems, such as conflicts between autonomy and paternalism in long-term care (Kane and Caplan, 1989). Specific examples include nursing home patients who do not wish to have their hours of rising and retiring, along with other activities, set by caregivers, or who want to be able to refuse therapy, even when it is indicated.

Issues of Equity and Access

- **The committee recommends research on issues of equity and access.**

The projected increase in the number of elderly Americans, the rising costs of Medicare, competing social needs, and continuing federal budget deficits have imposed artificial pressures on caregivers to limit health care spending in general and to limit care to the elderly in particular (as seen, for example, in the early discharge and decreased rehabilitative care given to patients operated on for hip fracture). Although difficult decisions regarding allocation of scarce resources are inevitable, no agreement now exists on how such rationing may be accomplished in an equitable manner that does not compromise access to basic health care resources (Daniels, 1986; Engelhardt and Rie, 1986; Miller and Miller, 1986).

Identification of Criteria for Appropriate Practice Standards

Standards must be developed for such health care services as cataract surgery, hip replacement, and cancer screening and treatment for the elderly. Given the wide regional variation in the use of such treatments, it is important to develop criteria that establish when these treatments become appropriate. Health services research on functional outcomes can elucidate the effectiveness of various therapies and identify those groups of patients for whom certain therapies are less effective. In addition, public forums and consensus conferences that focus on resource allocation can seek to forge public and professional agreement on the priorities to be set among various therapies to be made available to different patient populations.

Age as a Criterion for Allocating Health Care Resources

Chronological age per se has been proposed as the basis upon which health care expenditures should be limited (Brook, 1989). However, critics point out that the elderly are heterogeneous in both function and prognosis and that justice requires that costly therapies used in other age groups also should be scrutinized (Avorn, 1984; Callahan, 1987; Schneider, 1989). More data, debate, and analysis are needed to determine if or when health care resources should be allocated on the basis of age and to define those basic health care resources to which all elderly persons should have access. No data concerning how the American public regards rationed care based on age have been amassed.

It is important to identify misunderstandings of fact regarding rationing (e.g., that inordinate sums are spent on the intensive care of dying older patients) that might hinder policy decisions; educational programs must be devised to correct such misconceptions. Furthermore, the views of the public will be crucial in shaping a social consensus on these complicated and potentially divisive issues.

In addition to helping us determine how health care resources should be allocated, research can help us understand the current system of allocating health care services to the elderly (Avorn, 1984; Scitovsky and Capron, 1986; Anderson et al., 1989). For example, present incentives for early patient discharge, for outpatient in lieu of inpatient care, and for technological services rather than preventive and case management care may have particularly deleterious effects on the elderly (Grumet, 1989; Leaf, 1989).

The distribution of benefits and burdens of the current allocation system should be examined, discussed, and debated more explicitly to facilitate more realistic decision making.

Allocation of Resources for Patient Care and for Research

Resources available to devote to biomedical research, health services research, and patient care research are not unlimited. As suggested throughout this report, priorities need to be set to allocate these limited resources in a rational manner. Until recently, little research was dedicated to the most prevalent and disabling diseases of old age—the degenerative neurological diseases and the musculoskeletal diseases. Wide-ranging discussion, both with members of the public at large (especially the elderly) and with scientists, could help clarify the value choices that must be made to set future research priorities and could help establish procedures and criteria for allocating resources to the research endeavor.

Other research issues concerning allocation of resources include decisions about distribution of funds to support preventive care versus high-technology care in older persons, the allocation of funds to support health education compared with direct health care, the competition among different age cohorts for expensive treatment programs (e.g., bone marrow and organ transplant, hemodialysis, coronary artery bypass surgery), and the assignment of research support to study the role of poverty in health among elderly persons as compared to younger subjects.

Ethical Issues in Aging Research

- **The committee recommends research on biomedical ethical issues** involving participation in clinical research by frail elderly persons.

This report documents the need for significant increases in research on diseases and disorders afflicting the elderly. Disorders such as dementia, incontinence, and falls impair functioning and may lead to either home-based or institutional long-term care. Paradoxically, frail elderly persons with such disorders—ideal research subjects—may be among those least likely or physically able to participate as subjects of research (Brieger, 1978; Annas and Glantz, 1986b; Dubler, 1987; Kane and Manoukian, 1989). However, intensive research is needed to elucidate the etiology of these conditions and to find effective treatments. Patients with severe dementia lack the capacity to give informed consent to participate in research. Institutionalization, regardless of cognitive function, *impairs the autonomy of residents* , making it difficult for them to decline to participate in research (Cassel, 1985, 1987, 1988). These problems were amply documented by the National Commission for the Protection of Human Subjects of Biomedical and Behavioral Research in 1974 and led to appropriate restraint in the use of nursing home patients as research subjects (National Commission for the Protection of Human Subjects of Biomedical and Behavioral Research, 1978a, b). For this vitally needed research, the population of incapacitated and institutionalized people able to serve as research subjects, including those *not* of geriatric age, must be expanded. How best to accomplish this goal is, itself, an important research question.

Innovative Approaches to Informed Consent

Researchers should develop and evaluate the use of simplified or alternative ways to inform patients about research projects (Levine, 1986). For example, videotapes might enable persons with mild to moderate dementia to understand enough about a research project to give informed consent (Cassel, 1987). Similarly, linguistic analysis of impaired language processing might help physicians communicate better with research-eligible patients suffering from mild dementia.

Proxy Consent for Incompetent Patients

Consent by a family member or a close friend may not meet the standard of substituted judgment, which states that the proxy has given consent or refusal as desired by the incapacitated subject (Tymchuck et al., 1988). Studies have shown that proxies often give consent based on their own value system, not that of the patient.

Thus, the practice of obtaining consent from relatives of incapacitated patients needs to be reevaluated. Several methods to improve such substituted judgment should be considered, such as education of the public and, in particular, of relatives of patients eligible for research studies in the concept of proxy consent to research participation. Furthermore, educational interventions aimed at the elderly could encourage them to give advance directives for participating in research, just as they are encouraged to do for medical care decisions.

Institutional ethics committees might evaluate the appropriateness of a proxy's consent. In some situations, the appropriate standard for participation of incompetent patients in research might be that of the patient's best interests (Warren et al., 1986). Research and further discussion should disclose the specific situations in which the best-interest standard is most appropriate.

Attitudes Toward Participation in Research

Philosophers have posited that research should be a partnership between investigator and subject in which both parties understand the potential benefits to the patient and to the larger society (Fletcher et al., 1985; Veatch, 1987). However, some older people fear being research subjects (Faden and Beauchamp, 1986). Institutional safeguards are enacted based on the assumption that exploitation is possible, or even probable, without such protection. Thus, it becomes important to examine public attitudes about research. What responsibility does an ill person have to participate in research designed to understand and ameliorate conditions afflicting himself and others of a similar age? What discourages an elderly person from participating in research, and how might those barriers be overcome? Understanding how elderly people answer these questions would further our nation's sense of mission regarding research on aging.

Research Review in Long-Term Care

Although an increasing amount of research is being conducted in nursing homes, the process of review through institutional review boards (IRBs) remains sporadic and irregular (Ratzen, 1980). Nursing home and home care research affiliated with an academic center is generally reviewed by the center's IRB. However, such IRBs may not be knowledgeable about long- term care and may react in either an overprotective or an underprotective manner. Moreover, research lacking academic affiliation is being conducted increasingly in nursing homes and home care programs. In such cases, review of research may be ad hoc or totally lacking. In order to promote the quality of research, to educate investigators about the ethical dilemmas they face, and to protect research subjects, scientists should explore the establishment of community-based IRBs or other alternatives for review of methods used in research projects.

ADDITIONAL RESEARCH OPPORTUNITIES

The following topics provide additional research opportunities:

- treatment of older persons with experimental therapies that may involve increased risk to them, such as genetic engineering or organ transplants, which may raise ethical issues that require systematic study through research; and
- the trade-off between quality of care and quality of life for older patients.

CONCLUSIONS AND RESOURCE RECOMMENDATIONS

Ethical issues are an inevitable aspect of medical care and health care policy affecting the elderly. Furthermore, difficult ethical issues arise in the conduct of research on elderly subjects. However, to date, no federal resources have been allocated to conduct research on ethical issues concerning older persons. In order to carry out this important work, the federal government should apportion additional funds for research in biomedical ethics. An important step in this direction recently was taken by the National Institutes of Health's National Center for Human Genome Research, which earmarked 3 percent of its grant funds for study of ethical and social issues. Support for research on biomedical ethics in the treatment and study of older persons may help our society better understand and resolve these difficult dilemmas.

6

Research on Aging

It is difficult to identify the resources committed to research on aging. The limitations on measuring resources include the following: self-reporting by researchers; the absence of biologic markers to identify the research territory of old age; shared study targets, as in the examination of disorders found at several stages of life (e.g., diabetes, atherosclerosis, hypertension); the inclusion of administrative and training costs in research budgets; and the fact that basic discoveries in almost *any* area of biological science may contribute to the quality of life of older persons. Despite these limitations, estimates of current support and future needs for research on aging are needed in order to promote realistic planning for this expanding area of study.

FUNDING SUPPORT FOR RESEARCH ON AGING

The Federal Government

National Institutes of Health

Most federal support comes from the National Institutes of Health (NIH). In fiscal year 1990, funds committed to the study of basic biomedical aspects of aging were about $442 million, or 5.8 percent of all NIH research funds, with $239 million coming from the National Institute on Aging (NIA) (Office of Planning, Technology, Information and Evaluation, NIA). Table 7-1 summarizes funds for research on aging as reported by the institutes.

Department of Veterans Affairs

Veterans 65 and older constituted 27 percent of the veteran population in 1990, and this figure is expected to rise to 37 percent by the year 2000. The Department of Veterans Affairs (DVA) has been a leader in the development of research and training in geriatrics and gerontology in this country. In fiscal year 1990 the DVA reported about $17 million for age-related research, or 8.5 percent of the 1990 research budget of $201 million for the Department of Medicine and Surgery (Office of Assistant Chief Medical Director for Research and Development, DVA).

Alcohol, Drug Abuse, and Mental Health Administration

In fiscal year 1990 the Alcohol, Drug Abuse, and Mental Health Administration (ADAMHA) spent $42 million on research on aging, including $38 million on age-related research by the National Institute of Mental Health, or 4.9 percent of its total research budget of $855 million (Division of Financial Management, ADAMHA).

Health Care Financing Administration

The Health Care Financing Administration (HCFA) is the overseeing federal agency for Medicare and Medicaid, and in fiscal year 1990 it committed approximately $40 million to age-related studies, or about 82 percent of its research budget of $49 million for that year, largely in health services delivery (Office of Research and Demonstrations, HCFA). Most of this support was for demonstration projects; although demonstration projects can contribute to an understanding of aging, the committee believes that these funds cannot entirely be credited to support of research on aging.

Agency for Health Care Policy and Research

The Agency for Health Care Policy and Research (AHCPR)(formerly the National Center for Health Services Research and Health Care) spent $19 million on research on aging, or about 20 percent of its research budget of $95 million in fiscal year 1990 (Office of Financial Management, AHCPR). Most of the research supported was directed to the area of health services delivery.

Other Federal Departments and Agencies

Other agencies that provided support for research on aging included the Department of Agriculture, which committed $23 million in fiscal year 1990 for studies of nutrition in older persons (Office of Budget and Management Staff; Agriculture Research Service), and the Administration on Aging, which devoted about $3.0 million to research on aging that largely involved social and behavioral studies (Division of Research and Demonstrations) for the same year.

Industry and Foundations

Industry

Industry support for research on aging is difficult to measure because there is no central information repository. Indirect evidence comes from a review of the literature (see below), which showed that 5 percent of research papers on aging cited corporate support. A special case of research support by industry is that of the pharmaceutical companies. In 1989, U.S. companies spent more than $3.6 billion on research and development of drugs primarily used to treat diseases that afflict older patients (Pharmaceutical Manufacturers Association). This was 50 percent of the pharmaceutical industry's total research and development budget of $7.3 billion for the year. Cardiovascular disorders (stroke, heart disease, and hypertension) consumed 39 percent of the reported age-related research budget; the remainder of the budget supported investigation of drugs for the treatment of cancer, arthritis, and other conditions that afflict the geriatric population. Because most of these disorders also affect younger persons and because a significant portion of the funds went for development purposes, one cannot state with certainty how much of these funds actually supported age-related research.

Foundations

Information on foundation support for research on aging was obtained from the Foundation Center's 1989 publication *Grants for the Aged* and from a computer search of the center's on-line Grants Index. Both sources report information about corporate and foundation contributions of $5,000 or more. The Foundation Center reports that, from 1986 to early 1987, 2.5 percent of the monies given by private and corporate foundations were directed to the general area of aging, and less than 1 percent of all foundation philanthropy was for age-related research. In 1989, about $59 million was given for general projects in aging; of this sum, about $15 million (25 percent of all funds for projects in aging) went for studies specifically devoted to research on aging. The balance supported services, education, and administrative support for age-related programs. Foundations providing major support for research on aging included the Commonwealth Fund, New York; the Charles A. Dana Foundation, New York; the John A. Hartford Foundation, New York; the Henry J. Kaiser Family Foundation, California; the John D. and Catherine T. MacArthur Foundation, Illinois; the Robert Wood Johnson Foundation, New Jersey; the Weingart Foundation, California; and the Pew Charitable Trusts, Pennsylvania. Information about state and local support of age-related research is difficult to obtain. A few states, such as California (through the statewide Academic Geriatric Research and Education Program) provide limited funds to support studies on aging. Local (city and county) support for research on aging could not be estimated because of the absence of any central source for such information.

INSTITUTIONS ENGAGED IN RESEARCH ON AGING

Centers for Research on Aging Supported by the NIA

Centers for the Study of Alzheimer's Disease

The NIA has established centers for basic science and clinical studies in Alzheimer's disease at the following institutions:

- Baylor College of Medicine, Texas
- Case Western Reserve University, Ohio
- Columbia University, New York
- Duke University, North Carolina
- Harvard University, Massachusetts
- Johns Hopkins University, Maryland
- Mount Sinai School of Medicine, New York
- University of California at San Diego
- University of Kentucky
- University of Michigan
- University of Pittsburgh, Pennsylvania

- University of Southern California
- University of Texas Southwestern Medical Center
- University of Washington
- Washington University (St. Louis), Missouri

Centers of Excellence in Research and Teaching in Geriatrics and Gerontology (Claude Pepper Centers)

Three Claude Pepper centers have been funded by the NIA. The NIA also provides funds for research in numerous university-based research centers through program project grants (P01). However, it is not always possible to characterize the centers receiving such funds as primarily age-oriented in their research activities.

Geriatric Research, Education, and Clinical Centers of the Department of Veterans Affairs

There are 7 million veterans 65 years old or older; half of all men in this country who are 65 and older are veterans. By the year 2000 there will be 9 million veterans 65 and over; 4 million of these will be 75 and over. In response to the needs of the veteran population the Geriatric Research, Education, and Clinical Centers (GRECC) program was initiated by act of Congress in 1975, and was expanded by further legislation in 1980. There are now 13 GRECCs, with 3 more planned for fiscal year 1991. Their purpose is to integrate basic research, teaching, and clinical achievements. Research activities of these centers are summarized in Table 7-2.

Teaching and Research Nursing Homes

The Robert Wood Johnson Foundation has sponsored eight teaching and research nursing homes affiliated with university medical schools. Research activities at these institutions are summarized in Table 7–3.

Other Institutional Supports for the Study of Aging

A noteworthy step by the private sector towards recruitment for careers in geriatrics took place in 1988, as the John A. Hartford Foundation initiated a strategy of supporting centers of excellence. The program, enhanced and broadened during 1989, encourages medical students and practicing physicians at 10 sites across the

United States to specialize in academic geriatrics. Although there are dozens of university centers for the study of aging across the country (including biomedical, behavioral and social, and health services delivery programs), a 1987 review of medical school-based fellowships in geriatrics identified only 13, at that time, as having an adequate complement of research and teaching personnel and implementing a full research program in aging (Institute of Medicine, 1987).

TABLE: DVA Geriatric Research, Education, and Clinical Centers

Location	Research Activities
Ann Arbor, Mich.	Neurosciences; metabolic and endocrine factors; autonomic function in diabetes mellitus and hypertension
Bedford, Mass.	Neurobiology of aging; immunology; rheumatology; computer-based clinical evaluation
Brockton-West Roxbury, Mass.	Influence of normal aging on homeostatic mechanisms: cardiovascular, endocrine, and neuroendocrine; carbohydrate metabolism
Durham, N.C.	Cancer and cardiovascular disease in aging; research on physical fitness
Gainesville, Fla.	Geropharmacology; molecular biology of aging; gene expression and aging; nutrition; neoplasms
Little Rock, Ark.	General basic and clinical research; role of antioxidants; motor and cognitive deficits
Minneapolis, Minn.	Study of Parkinson's disease; Alzheimer's disease
Palo Alto, Calif.	Endocrinology; metabolism; age-related changes in cognition; approaches to health delivery; depression
San Antonio, Tex.	Metabolism; endocrinology; nutrition; oral health and dentistry
Seattle, Wash.	Age-related changes in behavior and neuroendocrine function, and their relationship to health services

Sepulveda, Calif.	Endocrinology; memory; health services delivery
St. Louis, Mo.	General basic and clinical oncology; hypertension; gastroenterology; infectious disease; nephrology
West Los	Angeles, Calif. Basic science and clinical study of immunology; osteoporosis; evaluation of health care

SOURCE: DVA Department of Medicine and Surgery. Office of Geriatrics and Extended Care, Washington, D.C.

PUBLICATIONS ON RESEARCH IN AGING

The scientific literature represents an important resource for all future research activities. In order to estimate changes in the knowledge base and to identify trends in research activity in geriatrics and gerontology, the Institute for Scientific Information (ISI) undertook a screen of 500,000 articles from 1983 to 1987. Funding supports and institutions of origin of scientific papers in 1982, 1983, and 1987 were assessed by ISI and by King Research, Inc., at the request of the Institute of Medicine.

In 1983, about 4,900 papers (or 1.7 percent of the total file of scientific reports) focused on aging; by 1987 the number of age-related research reports had risen to 8,900 (2.7 percent of the total file). The total file increased by 9 percent during this interval, resulting in an absolute increase in published studies on aging of 70 percent. Major areas of growth in the literature were in the neurosciences, including Alzheimer's disease, neurologic aging, cognition/ memory, and Parkinson's disease. Reports in the area of neurological function and Alzheimer's disease increased fourfold. Second in growth were reports on infectious disease and osteoporosis in the aged. Social and psychological studies, health services delivery research, and pharmacology were well represented in both years, showing a modest increase in the number of publications between 1983 and 1987. Examination of the social and behavioral literature showed a high level of interest in geriatrics in 1983, but no striking change in the rate of publication was noted in the subsequent four years. A further observation by the ISI review was that the increase in studies on aging involved many clearly defined areas of geriatric interest (e.g., studies of geriatric syndromes) with contributions made by many different fields of research (internal medicine, neurology, molecular biology, sociology, psychology). This observation demonstrates both the wide interest in the investigation of aging and the emergence of a body of information and research techniques that is focused on age-related topics.

TABLE: Areas of Study at Teaching and Research Nursing Homes

Institution	Research Activities
University of California, San Diego	Nosocomial infections; predictors of institutionalization; sleep apnea
Yeshiva University	Dementia; osteoarthritis; motor control impairment; age-associated memory failure; congestive heart failure
Harvard University	Syncope and altered blood pressure; homeostasis in the elderly; urinary incontinence; vitamin D physiology and nutrition; risk of institutionalization; Alzheimer's disease: depression, neuroendocrine function
Case Western Reserve University	Respiratory and gastrointestinal infections; immune processes and tuberculosis; visual perception and Alzheimer's disease; neuroendocrine function in Alzheimer's disease
Johns Hopkins University	Metabolic regulation; cardiopulmonary physiology; sleep physiology; neuropsychological function
University of Pennsylvania	Urinary tract infection; regional cerebral structure and function in dementia; sleep apnea
Stanford University	Modifiable factors influencing health status/ health care; hormones and hip fractures; cognitive function in diabetics; care in nursing homes
University of Iowa	Parkinson's disease; chronic conditions involving functional or cognitive impairment; interaction of stress and social supports

Until recently, as the ISI findings indicated, aging studies not reported in publications clearly identified as focused on aging have tended to appear in professional journals devoted to behavioral and social science or to health services delivery research, with few reports appearing in medical and other scientific publications.

That this trend has changed is illustrated, if not proven, by the increase between 1983 and 1987 in age-related studies reported in the selected medical and scientific journals shown in Tables.

Although these peer-reviewed journals showed a more than 150 percent increase in publication of articles on research in aging between 1983 and 1987, further study is necessary to establish the presence of a significant trend of interest is the quadrupling of articles on aging in the neurology literature.

During the period 1983 to 1987 there was also a significant increase in the number of journals devoted to the study of aging.

No support was listed for 16 percent of papers on age-related research in 1987; presumably, this support was provided by the institution of affiliation of the investigators. Among remaining papers, citations of support were divided as follows: government, 45 percent; foundations, 30 percent; foreign sources, 20 percent; and corporations, 5 percent.

PERSONNEL ENGAGED IN THE STUDY OF AGING

A 1980 study indicated that 7,000 to 10,300 geriatricians would be needed by the year 1990 (Kane et al., 1980). Based on staffing of 3 to 5 full-time faculty per teaching hospital and medical school, the study predicted a need for 2,100 biomedical faculty to provide training for these geriatricians.

It has been estimated that, by the year 2000, in order to meet the medical needs of the expanding older population, 10,000 to 21,000 geriatricians will be needed, and 14,000 to 29,000 will be needed by the year 2020 (NIA, 1988). Training of these physicians will require a major increase in staffing of departments of medicine and family practice by researchers and teachers with special competence in geriatrics and gerontology.

TABLE: Number of Age-Related Publications in Selected Journals

Journal	1983	1987
American Journal of Medicine	9	28
American Journal of Physiology	9	16
American Journal of Public Health	8	21
Annals of Internal Medicine	21	42
Annals of Neurology	17	54
Archives of Neurology	10	62
Brain Research	35	156
British Medical Journal	38	40
Journal of Clinical Investigation	5	9
Journal of Clinical Psychiatry	8	12
Journal of Comparative Neurology	20	72
Journal of Experimental Psychology	1	12
Journal of Neurochemistry	11	48
Journal of Neurology, Neuroscience and Psychiatry	12	41
Journal of Neuroscience	7	43
Journal of the American Medical Association	38	49
Lancet	36	56
Neurology	18	80
Neuropsychology	2	13
Neuroscience Letters1	6	63
New England Journal of Medicine	31	65
Proceedings, National Academy of Sciences	9	40
Science	13	32
Social Science and Medicine	3	23
Total	377	1,077

SOURCE: Institute for Scientific Information, Philadelphia, Pa.

In 1987 the Institute of Medicine reported that only 100 geriatric fellows were graduating per year from hospital-based medical school programs. The report estimated that to meet requirements for an estimated 2,100 faculty by the year 2000 to train future clinicians and biomedical students of aging, the number of graduating fellows should be increased by at least 100 percent (Institute of Medicine, 1987).

A report to Congress from the NIA (1988) and a study from the University of California at Los Angeles (Reuben, personal communication, 1990) support this recommendation.

Although a recent report (Reuben et al., 1990) indicates that despite a recent upsurge in interest in geriatrics by internists and family practitioners, with as many as 5,000 certified geriatricians in practice by the mid-1990s, this is far short of the estimated minimum of 10,000 required by that time to provide adequate care of the older population.

A survey in 1986 of 400 graduates of geriatric biomedical fellowship programs showed that two-thirds held academic positions, but only 35 percent of those surveyed committed more than 10 percent of their time to teaching and research (Siu et al., 1989).

The study pointed to a low rate of publication (1.3 papers per graduate annually) among the recent graduates compared with fellowship graduates in other disciplines, and it concluded that the goal of producing academic geriatricians is far from being met.

Thus, given even the most generous estimate of biomedical personnel engaged in research on aging, fewer than 150 graduates of fellowship programs in geriatrics spend more than 10 percent of their time in research.

Balancing this estimate of the small number of researchers engaged in studies of aging is the observation that 15,047 authors were listed in geriatric and gerontological publications (including behavioral and health services research studies) in 1987, an increase of 34 percent over 1982 (King Research, Inc.).

While this information cannot be interpreted as indicative of growth in the full-time complement of research investigators in aging, it points to a significant upward trend in the involvement of academic faculty in gerontologic and geriatric studies.

Biomedical age-related research is carried out by faculty M.D.s and biomedical Ph.D.s, with some studies in behavioral aspects of aging done by a small number of physician psychiatrists.

The reports on training cited above did not measure the need for scientists and practitioners in behavioral or social studies, or in health services research.

As noted in Chapters 4 and 5, support for training of Ph.D.s and other investigators in behavioral and social research or in health services delivery is inadequate, and the production of teachers and researchers in these fields must be expanded if *all* of the needs of the older generation are to be met. In a report to Congress (NIA, 1988) the NIA estimated the minimum requirements for faculty, including physicians, for the years 1990 and 7000.

The estimates included faculty in the areas of psychology, pharmacology, and nursing, but did not include information about faculty to train social scientists or other nonbiomedical researchers on aging. The estimates, totaling about 2,900 faculty for the year 1990, and about 6,000 for the year 2000, were based on the expectation that 25 percent of physician time and 50 percent of the time of other faculty would be committed to research.

TABLE: Faculty Needs in Age-Related Studies, 1990 and 2000

Faculty Work Place	1990	2000
Medical schools—physicians	600	1,300
Medical schools—other faculty	600	1,300
Nursing schools	750	1,500
Dental schools	80	120
Social work schools	300	1,000
Optometry schools	80	125
Pharmacy schools1	50	300
Clinical psychology schools	150	450

SOURCE: National Institute on Aging (1988).

A 1987 report stated that the numbers of geriatric and gerontological faculty —including teachers and scientists in the fields of behavioral and social studies, and health services delivery—were inadequate to provide adequate training for health care personnel to meet the medical and nonmedical needs of older persons (NIA, 1988).

In fiscal year 1990 the *total* number of trainees (biomedical; behavioral and social; neuroscience and neuropsychology) supported by the NIA was 425, a figure far short of the number required (Office of Planning, Technology, Information, and Evaluation, NIA). The training of faculty must be increased sharply and as soon as possible if the recommended minimum goal of 6,000 additional faculty (biomedical and nonbiomedical) for research and teaching in age-related studies and the care of older persons is to be reached by the year 2000.

7

Human Cloning: Pros and Cons

For the past five years, the prospect of human cloning has been the subject of considerable public attention and sharp moral debate, both in the United States and around the world. Since the announcement in February 1997 of the first successful cloning of a mammal (Dolly the sheep), several other species of mammals have been cloned. Although a cloned human child has yet to be born, and although the animal experiments have had low rates of success, the production of functioning mammalian cloned offspring suggests that the eventual cloning of humans must be considered a serious possibility.

In November 2001, American researchers claimed to have produced the first cloned human embryos, though they reportedly reached only a six-cell stage before they stopped dividing and died. In addition, several fertility specialists, both here and abroad, have announced their intention to clone human beings. The United States Congress has twice taken up the matter, in 1998 and again in 2001-2002, with the House of Representatives in July 2001 passing a strict ban on all human cloning, including the production of cloned human embryos. As of this writing, several cloning-related bills are under consideration in the Senate. Many other nations have banned human cloning, and the United Nations is considering an international convention on the subject. Finally, two major national reports have been issued on human reproductive cloning, one by the National Bioethics Advisory Commission (NBAC) in 1997, the other by the National Academy of Sciences (NAS) in January 2002. Both the NBAC and the NAS reports called for further consideration of the ethical and social questions raised by cloning.

The debate over human cloning became further complicated in 1998 when researchers were able, for the first time, to isolate human embryonic stem cells. Many scientists believe that these versatile cells, capable of becoming any type of cell in the body, hold great promise for understanding and treating many chronic diseases and conditions. Some scientists also believe that stem cells derived from cloned human embryos, produced explicitly for such research, might prove uniquely useful for studying many genetic diseases and devising novel therapies. Public reaction to the prospect of cloning-for-biomedical-research has been mixed: some Americans support it for its medical promise; others oppose it because it requires the exploitation and destruction of nascent human life, which would be created solely for research purposes.

Human Cloning: What Is at Stake?

The intense attention given to human cloning in both its potential uses, for reproduction as well as for research, strongly suggests that people do not regard it as just another new technology. Instead, we see it as something quite different, something that touches fundamental aspects of our humanity. The notion of cloning raises issues about identity and individuality, the meaning of having children, the difference between procreation and manufacture, and the relationship between the generations. It also raises new questions about the manipulation of some human beings for the benefit of others, the freedom and value of biomedical inquiry, our obligation to heal the sick (and its limits), and the respect and protection owed to nascent human life.

Finally, the legislative debates over human cloning raise large questions about the relationship between science and society, especially about whether society can or should exercise ethical and prudential control over biomedical technology and the conduct of biomedical research. Rarely has such a seemingly small innovation raised such big questions.

The Inquiry: Our Point of Departure

As Members of the President's Council on Bioethics, we have taken up the larger ethical and social inquiry called for in the NBAC and NAS reports, with the aim of advancing public understanding and informing public policy on the matter. We have attempted to consider human cloning (both for producing children and for biomedical research) within its larger human, technological, and ethical contexts, rather than to view it as an isolated technical development. We focus first on the broad human goods that it may serve as well as threaten, rather than on the immediate impact of the technique itself. By our broad approach, our starting on the plane of human goods, and our open spirit of inquiry, we hope to contribute to a richer and deeper understanding of what human cloning means, how we should think about it, and what we should do about it.

On some matters discussed in this report, Members of the Council are not of one mind. Rather than bury these differences in search of a spurious consensus, we have sought to present all views fully and fairly, while recording our agreements as well as our genuine diversity of perspectives, including our differences on the final recommendations to be made. By this means, we hope to help policymakers and the general public appreciate more thoroughly the difficulty of the issues and the competing goods that are at stake.

Fair and Accurate Terminology

There is today much confusion about the terms used to discuss human cloning, regarding both the activity involved and the entities that result. The Council stresses the importance of striving not only for accuracy but also for fairness, especially because the choice of terms can decisively affect the way questions are posed, and hence how answers are given. We have sought terminology that most accurately conveys the descriptive reality of the matter, in order that the moral arguments can then proceed on the merits. We have resisted the temptation to solve the moral questions by artful redefinition or by denying to some morally crucial element a name that makes clear that there is a moral question to be faced.

On the basis of (1) a careful analysis of the act of cloning, and its relation to the means by which it is accomplished and the purposes it may serve, and (2) an extensive critical examination of alternative terminologies, the Council has adopted the following definitions for the most important terms in the matter of human cloning:

Cloning: A form of reproduction in which offspring result not from the chance union of egg and sperm (sexual reproduction) but from the deliberate replication of the genetic makeup of another single individual (asexual reproduction).

Human cloning: The asexual production of a new human organism that is, at all stages of development, genetically virtually identical to a currently existing or previously existing human being. It would be accomplished by introducing the nuclear material of a human somatic cell (donor) into an oocyte (egg) whose own nucleus has been removed or inactivated, yielding a product that has a human genetic constitution virtually identical to the donor of the somatic cell. (This procedure is known as "somatic cell nuclear transfer," or SCNT). We have declined to use the terms "reproductive cloning" and "therapeutic cloning." We have chosen instead to use the following designations:

Cloning-to-produce-children: Production of a cloned human embryo, formed for the (proximate) purpose of initiating a pregnancy, with the (ultimate) goal of producing a child who will be genetically virtually identical to a currently existing or previously existing individual.

Cloning-for-biomedical-research: Production of a cloned human embryo, formed for the (proximate) purpose of using it in research or for extracting its stem cells, with the (ultimate) goals of gaining scientific knowledge of normal and abnormal development and of developing cures for human diseases.

Cloned human embryo: (a) A human embryo resulting from the nuclear transfer process (as contrasted with a human embryo arising from the union of egg and sperm). (b) The immediate (and developing) product of the initial act of cloning, accomplished by successful SCNT, whether used subsequently in attempts to produce children or in biomedical research.

SCIENTIFIC BACKGROUND

Cloning research and stem cell research are being actively investigated and the state of the science is changing rapidly; significant new developments could change some of the interpretations in our report. At present, however, a few general points may be highlighted.

The technique of cloning. The following steps have been used to produce live offspring in the mammalian species that have been successfully cloned. Obtain an egg cell from a female of a mammalian species. Remove its nuclear DNA, to produce an enucleated egg. Insert the nucleus of a donor adult cell into the enucleated egg, to produce a reconstructed egg. Activate the reconstructed egg with chemicals or electric current, to stimulate it to commence cell division. Sustain development of the cloned embryo to a suitable stage in vitro, and then transfer it to the uterus of a female host that has been suitably prepared to receive it. Bring to live birth a cloned animal that is genetically virtually identical (except for the mitochondrial DNA) to the animal that donated the adult cell nucleus.

Animal cloning: low success rates, high morbidity. At least seven species of mammals (none of them primates) have been successfully cloned to produce live births. Yet the production of live cloned offspring is rare and the failure rate is high: more than 90 percent of attempts to initiate a clonal pregnancy do not result in successful live birth. Moreover, the live-born cloned animals suffer high rates of deformity and disability, both at birth and later on. Some biologists attribute these failures to errors or incompleteness of epigenetic reprogramming of the somatic cell nucleus.

Attempts at human cloning. At this writing, it is uncertain whether anyone has attempted cloning-to-produce-children (although at least one physician is now claiming to have initiated several active clonal pregnancies, and others are reportedly working on it). We do not know whether a transferred cloned human embryo can progress all the way to live birth.

Stem cell research. Human embryonic stem cells have been isolated from embryos (produced by IVF) at the blastocyst stage or from the germinal tissue of fetuses. Human adult stem (or multipotent) cells have been isolated from a variety of tissues. Such cell populations can be differentiated in vitro into a number of different cell types, and are currently being studied intensely for their possible uses in regenerative medicine. Most scientists working in the field believe that stem cells (both embryonic and adult) hold great promise as routes toward cures and treatments for many human diseases and disabilities. All stem cell research is at a very early stage, and it is too soon to tell which approaches will prove most useful, and for which diseases.

The transplant rejection problem. To be effective as long-term treatments, cell transplantation therapies will have to overcome the immune rejection problem. Cells and tissues derived from adult stem cells and returned to the patient from whom they were taken would not be subject (at least in principle) to immune rejection.

Stem cells from cloned embryos. Human embryonic stem cell preparations could potentially be produced by using somatic cell nuclear transfer to produce a cloned human embryo, and then taking it apart at the blastocyst stage and isolating stem cells. These stem cells would be genetically virtually identical to cells from the nucleus donor, and thus could potentially be of great value in biomedical research. Very little work of this sort has been done to date in animals, and there are as yet no published reports of cloned human embryos grown to the blastocyst stage.

Although the promise of such research is at this time unknown, most researchers believe it will yield very useful and important knowledge, pointing toward new therapies and offering one of several possible routes to circumvent the immune rejection problem. Although some experimental results in animals are indeed encouraging, they also demonstrate some tendency even of cloned stem cells to stimulate an immune response.

The fate of embryos used in research. All extractions of stem cells from human embryos, cloned or not, involve the destruction of these embryos.

THE ETHICS OF CLONING-TO-PRODUCE-CHILDREN

Two separate national-level reports on human cloning (NBAC, 1997; NAS, 2002) concluded that attempts to clone a human being would be unethical at this time due to safety concerns and the likelihood of harm to those involved. The Council concurs in this conclusion. But we have extended the work of these distinguished bodies by undertaking a broad ethical examination of the merits of, and difficulties with, cloning-to-produce-children.

Cloning-to-produce-children might serve several purposes. It might allow infertile couples or others to have genetically-related children; permit couples at risk of conceiving a child with a genetic disease to avoid having an afflicted child; allow the bearing of a child who could become an ideal transplant donor for a particular patient in need; enable a parent to keep a living connection with a dead or dying child or spouse; or enable individuals or society to try to "replicate" individuals of great talent or beauty. These purposes have been defended by appeals to the goods of freedom, existence (as opposed to nonexistence), and well-being – all vitally important ideals.

A major weakness in these arguments supporting cloning-to-produce-children is that they overemphasize the freedom, desires, and control of parents, and pay insufficient attention to the well-being of the cloned child-to-be. The Council holds that, once the child-to-be is carefully considered, these arguments are not sufficient to overcome the powerful case against engaging in cloning-to-produce-children.

First, cloning-to-produce-children would violate the principles of the ethics of human research. Given the high rates of morbidity and mortality in the cloning of other mammals, we believe that cloning-to-produce-children would be extremely unsafe, and that attempts to produce a cloned child would be highly unethical. Indeed, our moral analysis of this matter leads us to conclude that this is not, as is sometimes implied, a merely temporary objection, easily removed by the improvement of technique. We offer reasons for believing that the safety risks might be enduring, and offer arguments in support of a strong conclusion: that conducting experiments in an effort to make cloning-to-produce-children less dangerous would itself be an unacceptable violation of the norms of research ethics. There seems to be no ethical way to try to discover whether cloning-to-produce-children can become safe, now or in the future.

If carefully considered, the concerns about safety also begin to reveal the ethical principles that should guide a broader assessment of cloning-to-produce-children: the principles of freedom, equality, and human dignity. To appreciate the broader human significance of cloning-to-produce-children, one needs first to reflect on the meaning of having children; the meaning of asexual, as opposed to sexual, reproduction; the importance of origins and genetic endowment for identity and sense of self; the meaning of exercising greater human control over the processes and "products" of human reproduction; and the difference between begetting and making. Reflecting on these topics, the Council has identified five categories of concern regarding cloning-to-produce-children. (Different Council Members give varying moral weight to these different concerns.)

Problems of identity and individuality. Cloned children may experience serious problems of identity both because each will be genetically virtually identical to a human being who has already lived and because the expectations for their lives may be shadowed by constant comparisons to the life of the "original."

Concerns regarding manufacture. Cloned children would be the first human beings whose entire genetic makeup is selected in advance. They might come to be considered more like products of a designed manufacturing process than "gifts" whom their parents are prepared to accept as they are. Such an attitude toward children could also contribute to increased commercialization and industrialization of human procreation.

The prospect of a new eugenics. Cloning, if successful, might serve the ends of privately pursued eugenic enhancement, either by avoiding the genetic defects that may arise when human reproduction is left to chance, or by preserving and perpetuating outstanding genetic traits, including the possibility, someday in the future, of using cloning to perpetuate genetically engineered enhancements.

Troubled family relations. By confounding and transgressing the natural boundaries between generations, cloning could strain the social ties between them. Fathers could become "twin brothers" to their "sons"; mothers could give birth to their genetic twins; and grandparents would also be the "genetic parents" of their grandchildren. Genetic relation to only one parent might produce special difficulties for family life.

Effects on society. Cloning-to-produce-children would affect not only the direct participants but also the entire society that allows or supports this activity. Even if practiced on a small scale, it could affect the way society looks at children and set a precedent for future nontherapeutic interventions into the human genetic endowment or novel forms of control by one generation over the next. In the absence of wisdom regarding these matters, prudence dictates caution and restraint.

Conclusion: For some or all of these reasons, the Council is in full agreement that cloning-to-produce-children is not only unsafe but also morally unacceptable, and ought not to be attempted.

THE ETHICS OF CLONING-FOR-BIOMEDICAL-RESEARCH

Ethical assessment of cloning-for-biomedical-research is far more vexing. On the one hand, such research could lead to important knowledge about human embryological development and gene action, both normal and abnormal, ultimately resulting in treatments and cures for many dreaded illnesses and disabilities. On the other hand, the research is morally controversial because it involves the deliberate production, use, and ultimate destruction of cloned human embryos, and because the cloned embryos produced for research are no different from those that could be implanted in attempts to produce cloned children. The difficulty is compounded by what are, for now, unanswerable questions as to whether the research will in fact yield the benefits hoped for, and whether other promising and morally nonproblematic approaches might yield comparable benefits. The Council, reflecting the differences of opinion in American society, is divided regarding the ethics of research involving (cloned) embryos. *Yet we agree that all parties to the debate have concerns vital to defend, vital not only to themselves but to all of us. No human being and no society can afford to be callous to the needs of suffering humanity, or cavalier about the treatment of nascent human life, or indifferent to the social effects of adopting one course of action rather than another.*

To make clear to all what is at stake in the decision, Council Members have presented, as strongly as possible, the competing ethical cases for and against cloning-for-biomedical-research in the form of first-person attempts at moral suasion. Each case has tried to address what is owed to suffering humanity, to the human embryo, and to the broader society. Within each case, supporters of the position in question speak only for themselves, and not for the Council as a whole.

A. The Moral Case for Cloning-for-Biomedical-Research

The moral case for cloning-for-biomedical-research rests on our obligation to try to relieve human suffering, an obligation that falls most powerfully on medical practitioners and biomedical researchers. We who support cloning-for-biomedical-research all agree that it may offer uniquely useful ways of investigating and possibly treating many chronic debilitating diseases and disabilities, providing aid and relief to millions. We also believe that the moral objections to this research are outweighed by the great good that may come from it. Up to this point, we who support this research all agree. But we differ among ourselves regarding the weight of the moral objections, owing to differences about the moral status of the cloned embryo. These differences of opinion are sufficient to warrant distinguishing two different moral positions within the moral case for cloning-for-biomedical-research:

Position Number One. Most Council Members who favor cloning-for-biomedical-research do so with serious moral concerns. Speaking only for ourselves, we acknowledge the following difficulties, but think that they can be addressed by setting proper boundaries.

Intermediate moral status. While we take seriously concerns about the treatment of nascent human life, we believe there are sound moral reasons for not regarding the embryo in its earliest stages as the moral equivalent of a human person. We believe the embryo has a developing and intermediate moral worth that commands our special respect, but that it is morally permissible to use early-stage cloned human embryos in important research under strict regulation.

Deliberate creation for use. We believe that concerns over the problem of deliberate creation of cloned embryos for use in research have merit, but when properly understood should not preclude cloning-for-biomedical-research. These embryos would not be "created for destruction," but for use in the service of life and medicine. They would be destroyed in the service of a great good, and this should not be obscured.

Going too far. We acknowledge the concern that some researchers might seek to develop cloned embryos beyond the blastocyst stage, and for those of us who believe that the cloned embryo has a developing and intermediate moral status, this is a very real worry. We approve, therefore, only of research on cloned embryos that is strictly limited to the first fourteen days of development – a point near when the primitive streak is formed and before organ differentiation occurs.

Other moral hazards. We believe that concerns about the exploitation of women and about the risk that cloning-for-biomedical-research could lead to cloning-to-produce-children can be adequately addressed by appropriate rules and regulations. These concerns need not frighten us into abandoning an important avenue of research.

Position Number Two. A few Council Members who favor cloning-for-biomedical-research do not share all the ethical qualms expressed above. Speaking only for ourselves, we hold that this research, at least for the purposes presently contemplated, presents no special moral problems, and therefore should be endorsed with enthusiasm as a potential new means of gaining knowledge to serve humankind. Because we accord no special moral status to the early-stage cloned embryo and believe it should be treated essentially like all other human cells, we believe that the moral issues involved in this research are no different from those that accompany any biomedical research. What is required is the usual commitment to high standards for the quality of research, scientific integrity, and the need to obtain informed consent from donors of the eggs and somatic cells used in nuclear transfer.

B. The Moral Case against Cloning-for-Biomedical-Research

The moral case against cloning-for-biomedical-research acknowledges the possibility – though purely speculative at the moment – that medical benefits might come from this particular avenue of experimentation. But we believe it is morally wrong to exploit and destroy developing human life, even for good reasons, and that it is unwise to open the door to the many undesirable consequences that are likely to result from this research. We find it disquieting, even somewhat ignoble, to treat what are in fact seeds of the next generation as mere raw material for satisfying the needs of our own. Only for very serious reasons should progress toward increased knowledge and medical advances be slowed. But we believe that in this case such reasons are apparent.

Moral status of the cloned embryo. We hold that the case for treating the early-stage embryo as simply the moral equivalent of all other human cells (Position Number Two, above) is simply mistaken: it denies the continuous history of human individuals from the embryonic to fetal to infant stages of existence; it misunderstands the meaning of potentiality; and it ignores the hazardous moral precedent that the routinized creation, use, and destruction of nascent human life would establish.

We hold that the case for according the human embryo "intermediate and developing moral status" (Position Number One, above) is also unconvincing, for reasons both biological and moral. Attempts to ground the limited measure of respect owed to a maturing embryo in certain of its developmental features do not succeed, and the invoking of a "special respect" owed to nascent human life seems to have little or no operative meaning if cloned embryos may be created in bulk and used routinely with impunity.

If from one perspective the view that the embryo seems to amount to little may invite a weakening of our respect, from another perspective its seeming insignificance should awaken in us a sense of shared humanity and a special obligation to protect it.

The exploitation of developing human life. To engage in cloning-for-biomedical-research requires the irreversible crossing of a very significant moral boundary: the creation of human life expressly and exclusively for the purpose of its use in research, research that necessarily involves its deliberate destruction. If we permit this research to proceed, we will effectively be endorsing the complete transformation of nascent human life into nothing more than a resource or a tool. Doing so would coarsen our moral sensibilities and make us a different society: one less humble toward that which we cannot fully understand, less willing to extend the boundaries of human respect ever outward, and more willing to transgress moral boundaries once it appears to be in our own interests to do so.

Moral harm to society. Even those who are uncertain about the precise moral status of the human embryo have sound ethical-prudential reasons to oppose cloning-for-biomedical-research. Giving moral approval to such research risks significant moral harm to our society by (1) crossing the boundary from sexual to asexual reproduction, thus approving in principle the genetic manipulation and control of nascent human life; (2) opening the door to other moral hazards, such as cloning-to-produce-children or research on later-stage human embryos and fetuses; and (3) potentially putting the federal government in the novel and unsavory position of mandating the destruction of nascent human life. Because we are concerned not only with the fate of the cloned embryos but also with where this research will lead our society, we think prudence requires us not to engage in this research.

What we owe the suffering. We are certainly not deaf to the voices of suffering patients; after all, each of us already shares or will share in the hardships of mortal life. We and our loved ones are all patients or potential patients. But we are not only patients, and easing suffering is not our only moral obligation. As much as we wish to alleviate suffering now and to leave our children a world where suffering can be more effectively relieved, we also want to leave them a world in which we and they want to live – a world that honors moral limits, that respects all life whether strong or weak, and that refuses to secure the good of some human beings by sacrificing the lives of others.

PUBLIC POLICY OPTIONS

The Council recognizes the challenges and risks of moving from moral assessment to public policy. Reflections on the "social contract" between science and society highlight both the importance of scientific freedom and the need for boundaries. We note that other countries often treat human cloning in the context of a broad area of biomedical technology, at the intersection of reproductive technology, embryo research, and genetics, while the public policy debate in the United States has treated cloning largely on its own. We recognize the special difficulty in formulating sound public policy in this area, given that the two ethically distinct matters-cloning-to-produce-children and cloning-for-biomedical-research-will be mutually affected or implicated in any attempts to legislate about either. Nevertheless, our ethical and policy analysis leads us to the conclusion that some deliberate public policy at the federal level is needed in the area of human cloning.

We reviewed the following seven possible policy options and considered their relative strengths and weaknesses: (1) Professional self-regulation but no federal legislative action ("self-regulation"); (2) A ban on cloning-to-produce-children, with neither endorsement nor restriction of cloning-for-biomedical-research ("ban plus silence"); (3) A ban on cloning-to-produce-children, with regulation of the use of cloned embryos for biomedical research ("ban plus regulation"); (4) Governmental regulation, with no legislative prohibitions ("regulation of both"); (5) A ban on all human cloning, whether to produce children or for biomedical research ("ban on both"); (6) A ban on cloning-to-produce-children, with a moratorium or temporary ban on cloning-for-biomedical-research ("ban plus moratorium"); or (7) A moratorium or temporary ban on all human cloning, whether to produce children or for biomedical research ("moratorium on both").

THE COUNCIL'S POLICY RECOMMENDATIONS

Having considered the benefits and drawbacks of each of these options, and taken into account our discussions and reflections throughout this report, the Council recommends two possible policy alternatives, each supported by a portion of the Members.

Majority Recommendation: Ten Members of the Council recommend *a ban on cloning-to-produce-children combined with a four-year moratorium on cloning-for-biomedical-research. We also call for a federal review of current and projected practices of human embryo research, pre-implantation genetic diagnosis, genetic modification of human embryos and gametes, and related matters, with a view to recommending and shaping ethically sound policies for the entire field.*

Speaking only for ourselves, those of us who support this recommendation do so for some or all of the following reasons:

- By permanently banning cloning-to-produce-children, this policy gives force to the strong ethical verdict against cloning-to-produce-children, unanimous in this Council (and in Congress) and widely supported by the American people. And by enacting a four-year moratorium on the creation of cloned embryos, it establishes an additional safeguard not afforded by policies that would allow the production of cloned embryos to proceed without delay.
- It calls for and provides time for further democratic deliberation about cloning-for-biomedical research, a subject about which the nation is divided and where there remains great uncertainty. A national discourse on this subject has not yet taken place in full, and a moratorium, by making it impossible for either side to cling to the status-quo, would force both to make their full case before the public. By banning all cloning for a time, it allows us to seek moral consensus on whether or not we should cross a major moral boundary (creating nascent cloned human life solely for research) and prevents our crossing it without deliberate decision. It would afford time for scientific evidence, now sorely lacking, to be gathered – from animal models and other avenues of human research – that might give us a better sense of whether cloning-for-biomedical-research would work as promised, and whether other morally nonproblematic approaches might be available. It would promote a fuller and better-informed public debate. And it would show respect for the deep moral concerns of the large number of Americans who have serious ethical objections to this research.
- Some of us hold that cloning-for-biomedical-research can never be ethically pursued, and endorse a moratorium to enable us to continue to make our case in a democratic way. Others of us support the

moratorium because it would provide the time and incentive required to develop a system of national regulation that might come into use if, at the end of the four-year period, the moratorium were not reinstated or made permanent. Such a system could not be developed overnight, and therefore even those who support the research but want it regulated should see that at the very least a pause is required. In the absence of a moratorium, few proponents of the research would have much incentive to institute an effective regulatory system. Moreover, the very process of proposing such regulations would clarify the moral and prudential judgments involved in deciding whether and how to proceed with this research.

- A moratorium on cloning-for-biomedical-research would enable us to consider this activity in the larger context of research and technology in the areas of developmental biology, embryo research, and genetics, and to pursue a more comprehensive federal regulatory system for setting and executing policy in the entire area.
- Finally, we believe that a moratorium, rather than a lasting ban, signals a high regard for the value of biomedical research and an enduring concern for patients and families whose suffering such research may help alleviate. It would reaffirm the principle that science can progress while upholding the community's moral norms, and would therefore reaffirm the community's moral support for science and biomedical technology.

The decision before us is of great importance. Creating cloned embryos for *any* purpose requires crossing a major moral boundary, with grave risks and likely harms, and once we cross it there will be no turning back. Our society should take the time to make a judgment that is well-informed and morally sound, respectful of strongly held views, and representative of the priorities and principles of the American people. We believe this ban-plus-moratorium proposal offers the best means of achieving these goals. This position is supported by Council Members Rebecca S. Dresser, Francis Fukuyama, Robert P. George, Mary Ann Glendon, Alfonso Gómez-Lobo, William B. Hurlbut, Leon R. Kass, Charles Krauthammer, Paul McHugh, and Gilbert C. Meilaender.

Minority Recommendation: Seven Members of the Council recommend *a ban on cloning-to-produce-children, with regulation of the use of cloned embryos for biomedical research.* Speaking only for ourselves, those of us who support this recommendation do so for some or all of the following reasons:

- By permanently banning cloning-to-produce-children, this policy gives force to the strong ethical verdict against cloning-to-produce-children, unanimous in this Council (and in Congress) and widely supported by the American people. We believe that a ban on the transfer of cloned embryos to a woman's uterus would be a sufficient and effective legal safeguard against the practice.
- *It approves cloning-for-biomedical-research and permits it to proceed without substantial delay.* This is the most important advantage of this proposal. The research shows great promise, and its actual value can only be determined by allowing it to go forward now. Regardless of how much time we allow it, no amount of experimentation with animal models can provide the needed understanding of human diseases. The special benefits from working with stem cells from cloned human embryos cannot be obtained using embryos obtained by IVF. We believe this research could provide relief to millions of Americans, and that the government should therefore support it, within sensible limits imposed by regulation.
- It would establish, *as a condition of proceeding,* the necessary regulatory protections to avoid abuses and misuses of cloned embryos. These regulations might touch on the secure handling of embryos, licensing and prior review of research projects, the protection of egg donors, and the provision of equal access to benefits.
- Some of us also believe that mechanisms to regulate cloning-for-biomedical-research should be part of a larger regulatory program governing all research involving

> human embryos, and that the federal government should initiate a review of present and projected practices of human embryo research, with the aim of establishing reasonable policies on the matter.

Permitting cloning-for-biomedical-research now, while governing it through a prudent and sensible regulatory regime, is the most appropriate way to allow important research to proceed while insuring that abuses are prevented. We believe that the legitimate concerns about human cloning expressed throughout this report are sufficiently addressed by this ban-plus-regulation proposal, and that the nation should affirm and support the responsible effort to find treatments and cures that might help many who are suffering.

8

Human Cloning: An Overview

The prospect of human cloning burst into the public consciousness in 1997, following the announcement of the successful cloning of Dolly the sheep. It has since captured much attention and generated great debate, both in the United States and around the world. Many are repelled by the idea of producing children who would be genetically virtually identical to pre-existing individuals, and believe such a practice unethical. But some see in such cloning the possibility to do good for infertile couples and the broader society. Some want to outlaw it, and many nations have done so. Others believe the benefits outweigh the risks and the moral concerns, or they oppose legislative interference with science and technology in the name of freedom and progress.

Complicating the national dialogue about human cloning is the isolation in 1998 of human embryonic stem cells, which many scientists believe to hold great promise for understanding and treating many chronic diseases and conditions. Some scientists also believe that stem cells derived from cloned human embryos, produced explicitly for such research, might prove to be uniquely useful for studying many genetic diseases and devising novel therapies. Public reaction to this prospect has been mixed, with some Americans supporting it in the hope of advancing biomedical research and helping the sick and the suffering, while others are concerned about the instrumentalization or abuse of nascent human life and the resulting danger of moral insensitivity and degradation.

In the United States, several attempts have been made to initiate a comprehensive public review of the significance of human cloning and to formulate appropriate policies. Most notably, the

National Bioethics Advisory Commission (NBAC) released a report on the subject of cloning-to-produce-children in 1997. The Commission concluded that cloning-to-produce-children was, at least for the time being, unethical on safety grounds, and that the deeper and more permanent moral concerns surrounding the practice should be the subject of continuing deliberation "in order to further our understanding of the ethical and social implications of this technology and to enable society to produce appropriate long-term policies regarding this technology" (p. 106).

In this report, the President's Council on Bioethics takes up this important charge, and considers the ethical and social implications of human cloning (both for producing children and for biomedical research) in their full scope, with the aim of informing public policy on the matter.

Our work toward this end is guided by a number of explicit methodological choices about modes of approach, points of departure, and spirit of inquiry. We locate human cloning within its larger human and technological context, rather than consider it in isolation. We focus first on the broad human goods that it may serve or threaten, rather than on the immediate impact of the technique itself. And we present the strongest arguments for the relevant moral and policy positions, rather than frame the arguments in order to seek consensus. By our broad approach, our starting on the plane of human goods, and our open spirit of inquiry, we hope to contribute to a richer and deeper understanding of what human cloning entails, how we should think about it, and what we should do about it.

Two points of clarification before we proceed. First, all of our considerations and arguments assume that cloning techniques, both for producing children and for providing embryos useful in biomedical research, could succeed in human beings as they have with other mammals. Cloning-to-produce-children has never been successfully carried out in humans, and cloning embryos for biomedical research has not progressed beyond the earliest experiments. We consider it part of our task to judge whether even attempts at human cloning would be ethical or should be lawful. To conduct the analysis and assessment needed for such judgment, we necessarily proceed on the assumption, which we believe is supported by evidence from animal experiments, that human cloning is indeed a possibility – that sooner or later, if it were allowed and attempted, human cloning could be successfully carried out. Practically all public discussion of the ethics of human cloning has, whether expressly or not, proceeded on this same premise, and rightly so.

Second, on some of the matters discussed in this report, Members of the Council are not of one mind. Given that competing goods are at stake, and different people regard them differently, this is not at all surprising. Rather than bury these differences in search of a spurious consensus, we have sought to present all views fairly and fully. Yet transcending these differences is a more fundamental agreement about the worthiness of the approach we have adopted and the arguments we have made. Accordingly, the Council is unanimous in owning the entire report and in recommending, to all, the report's discussions and arguments for serious consideration.

In the remainder of this overview, we describe the context of human cloning and the discussions it has generated. In the course of doing so, we identify the kinds of questions and concerns that would permit a full assessment of the meaning of human cloning. These questions and concerns will guide us throughout the report.

HUMAN CLONING IN CONTEXT

It is useful to begin by observing how it is that the question of human cloning has come before us. The prospect of cloning human beings confronts us now not as the result of a strong public demand or a long-standing need. Unlike sought-for medical therapies, it was not at the outset pursued as a cure for disease. Neither has it been sought explicitly as a tool for genetic control or "enhancement" of human offspring. Cloning has arisen not so much because it was actively sought for its own sake, but because it is a natural extension of certain biotechnological advances of the past several decades.

For more than half a century, and at an accelerating pace, biomedical scientists have been gaining wondrous new knowledge of the workings of living beings, from small to great. Increasingly, they also are providing precise and sophisticated knowledge of the workings of the human body and mind. Such knowledge of how things work often leads to new technological powers to control or alter these workings, powers ordinarily sought in order to treat human disease and relieve suffering.

Questions regarding the meaning of acquiring such powers – both the promise and the peril – have attracted scholarly and public attention. For more than thirty years, ethical issues related to biomedical advance have occupied the growing field of bioethics. Increasingly, these ethical issues have spawned public discussion and debates. A growing number of people sense that something new and momentous is happening; that the accelerating waves of biotechnical advances touch deeply on our most human concerns; and that the centuries-old project for human mastery of nature may now be, so to speak, coming home, giving humanity the power to alter and "master" itself.

One important aspect of human life already affected by new biotechnologies is human reproduction. For several decades now, building on advances in genetics, cell biology, and developmental biology, and on technologies used first in animal husbandry, scientists around the world have been adapting techniques and developing tools to study, influence, and manipulate the origins of human life. Beginning with techniques of artificial insemination and progressing through in vitro fertilization (IVF) and intracytoplasmic sperm injection, artificial aids to reproduction have come into standard medical use.

Cloning is, in one sense, another step along this path. It developed as the result of research into mammalian reproduction and development, where it is desired also as a means of replicating animals especially useful to human beings. It is also proposed as an additional means to overcome infertility in humans.

But the controversy surrounding human cloning, and the widespread sense of disquiet and concern with which the prospect has been received around the world, make it clear that cloning is not just another reproductive technology, to be easily assimilated into ordinary life. Nearly all participants in the public debate over human cloning appear to agree that the subject touches upon some of the most fundamental questions regarding the nature of our humanity and the character of our society. In addition, it raises questions about the aims of biomedical science and about the relation between science and society, including the possibility and desirability of exercising public control over the uses of biomedical technology and the conduct of biomedical research. It is because we sense these larger entailments that the subject of cloning matters so much to us. It is these considerations that give the present debate its force and prominence. Thus only through a serious reflection on these broader questions can the full meaning of human cloning be discovered. The prospect of human cloning may have been brought before us by the march of biotechnology, but now that it is here it is incumbent upon us to look well beyond its technical and medical aspects, if we are to appreciate its significance in full.

Three areas of inquiry in particular seem essential to any understanding of the full meaning of human cloning: the nature and meaning of human procreation; the aims, ends, and means of biomedical science and technology; and the relation of science and technology to the larger society.

CLONING AND HUMAN PROCREATION

Human procreation provides the major context for considering the prospect of cloning, especially cloning-to-produce-children. Much of the time, most of us tend to take for granted this central aspect of human life, through which all of us come to be and through which we give birth to our posterity. But the prospect of creating children by cloning brings this subject sharply before us and compels us to examine the nature and meaning of human procreation. For cloning-to-produce-children, while it may be a potential aid to human reproduction, appears also to be a substitute for it, or at least for its natural, un-programmed, sexual character. Properly to assess the meaning of producing cloned children, one must first of all consider the meaning of human procreation in all its aspects and entailments.

Human procreation, though seemingly an exclusively private act, has a profoundly public meaning. It determines the relations between one generation and the next, shapes identities, creates attachments, and sets up responsibilities for the care and rearing of children (and the care of aging parents or other needy kin).

Thus, in considering proposals to clone children, we must ask ourselves what cloning would mean not only for the individual parents and children involved, but also for the surrounding families and for all of society. Opinions on this subject will of course differ, sometimes widely, as they rest on possibly differing perceptions of human procreation and family life. Yet the following basic observations, concerns, and questions seem pertinent, notwithstanding possible differences of opinion among us about how much weight to give them.

Among the important aspects of the topic are these: the meaning of having children; the meaning of sexual, as opposed to asexual, reproduction; the meaning of origins and genetic endowment for identity and sense of self; the meaning of exercising greater human control over the processes and "products" of human reproduction; and the difference between begetting and making.

To understand what it would mean to clone a child, we do well to consider most generally what it means to bring a child into the world, and with what attitude we should regard his or her arrival and presence. Our children are, to begin with, our replacements, those who will one day stand in our place. They are, as Hans Jonas has remarked, "life's own answer to mortality." Though their conception is the fruit of our activity, and though we are responsible for saying "yes" to their arrival, we do not, in normal procreation, command their conception, control their makeup, or rule over their development and birth. They are, in an important sense, "given" to us. Though they are *our* children, they are not our *property*. Though they are our flesh and blood, and deeply kin, they are also independent "strangers" who arrive suddenly out of the darkness and whom we must struggle to get to know. Though we may seek to have them for our own self-fulfillment, they exist also and especially for their own sakes. Though we seek to educate them, they are not like our other projects, determined strictly according to our plans and serving only our desires.

If these observations are correct, certain things follow regarding the attitudes we should have toward our children. We treat them rightly when we treat them as gifts rather than as products, and when we treat them as independent beings whom we are duty-bound to protect and nurture rather than as extensions of ourselves subject only to our wills and whims. Might these attitudes toward children be altered by cloning, and, if so, how? Would social attitudes toward children change, even if cloning were not practiced widely? What might these changes mean?

To understand how the introduction of *a*sexual reproduction might affect human life, we must first seek the intrinsic meaning of the *sexual* character of human reproduction and what it implies for individuals, for families, and for the relation between the generations. Once again, the following observations – while hardly exhaustive – seem pertinent and important.

In sexual reproduction, each child has two complementary biological progenitors. Each child thus stems from and unites exactly two lineages, lines that trace backward in similar branching fashion for ages. Moreover, the precise genetic endowment of each child is determined by a combination of nature and chance, not by human design: each human child naturally acquires and shares the common human species genotype, each child is genetically (equally) kin to each (both) parent(s), yet each child is also genetically unique. Cloning-to-produce-children departs from this pattern. A cloned child has unilineal, not bilineal, descent; he or she is genetically kin to only one progenitor. What is more, the genetic kinship is near-total: the cloned child is not genetically unique, but shares almost completely the genetic endowment of the "original" progenitor. Finally, this endowment comes to the cloned child not by chance but by human choice and decision. What do these differences mean for the cloned child, for family relations, and for relations across the generations?

Origins and genetic endowment are significant aspects of who one is and how one regards oneself, of one's "identity," individuality, and place in the social order. The biological linkages and prospects implicit in sexual reproduction help to define us, though, it should go without saying, they do not define us completely. While we are more "what we choose to become" than we are "where we came from," our human beginnings matter, biologically, psychically, and socially. Because of the way we are generated, each of us is at once (1) equally human, (2) equally marked by and from birth as mortal, (3) equally enmeshed in a particular familial nexus of origin, (4) equally individuated in our trajectory from the beginning to the end of our lives – and, if all goes well, (5) equally capable (despite our mortality) of participating with a complementary other in the very same renewal of human possibility through procreation. Our genetic identity – manifest, for instance, in our distinctive appearance by which we are recognized by others and in our immune system by which we maintain our integrity against "foreign invasions" – also symbolizes and foreshadows exactly the unique, never-to-be-repeated character of each human life. In addition, human societies virtually everywhere have structured child-rearing responsibilities and systems of identity and relationships on the bases of these natural facts of begetting.

Kinship is tied to origins, and identity, at least in part, is tied to kinship. It is against this background that we must consider the implications of clonal reproduction, and the alterations it might produce in how cloned children would regard themselves and how they would be regarded by others. What would cloning-to-produce-children mean for individual identity, for kinship, and for sense of self, not only for the cloned child but also for his or her family?

Unaided sexual procreation is an activity at once natural, private, mysterious, unmediated, unpredictable, and undesigned. With the arrival of techniques such as IVF to assist procreation in the face of infertility, the process becomes less private and more mediated. But although technique is used, the basic structure of sexual reproduction – the combination of genetic material from father and mother resulting in a genetically unique child – is unaltered, the outcome is still unpredictable, and the genetic endowment of the child remains uncontrolled and undesigned. Cloning-to-produce-children would seem to bring procreation under human control and direction. What would this mean? What are the implications of allowing reproductive activities to become increasingly technological and commercialized? Cloning would be the first instance in which parents could select in advance the precise (or nearly precise) genetic makeup of their child, by selecting the donor to be cloned. It therefore forces us to ask what might be the difference between begetting and making, to wonder whether cloning somehow crosses the line between them, and, if so, to consider whether and why that should worry us.

Though admittedly sketchy and incomplete, these preliminary reflections on the nature and meaning of human procreation should enable us to see cloning – and especially cloning-to-produce-children-in its most important human context and to understand its deepest implications for its practitioners and for society.

CLONING AND BIOMEDICAL SCIENCE

Human procreation is not the only context for evaluating the prospect of human cloning. As a product of biotechnology, a potential means of assisted reproduction, and a possible source of cloned embryos for research and medical use, human cloning also points us to questions about the aims, ends, and means of biomedical science and technology. Ordinarily, we are not prompted to much reflection about what science is for and what goals technology should serve. Our society tacitly accepts the self-directing and self-augmenting character of these activities, and the vast majority of us support them because we esteem and benefit from their contributions to human understanding and human welfare. However, when developments such as cloning raise profound questions affecting fundamental moral values and social institutions, we are forced to consider the ends and means of science and technology, and to explore their standing in the scheme of human goods.

To provide a context for assessing human cloning and its possible benefits, we do well to remember the goals of medicine and modern science: the great value and importance of treating disease and relieving suffering, including the sorrows of infertility; and the great value and importance of gaining knowledge about the workings of nature, our own nature emphatically included.

No one can doubt the merit of these noble aims. Yet there has always been some disagreement about the lengths to which we should allow ourselves to go in serving them. Questions therefore arise about the need for limits on scientific pursuits and technological activities, and, conversely, about the meaning of such limits for the scientific and technological enterprises.

To address these questions, we must appreciate the human good of biomedical science in its fullness, and we must ask about the necessary and sufficient conditions for its flourishing. We must recognize, among other things, the unpredictability of scientific discovery and technological innovation, and the importance, therefore, of keeping open lines of inquiry and experimentation regardless of current estimates of their likelihood of success. Although serendipity often favors the prepared mind, nature guards her secrets well, and even the best scientists are regularly surprised by where the keys to the locks are ultimately found.

But precisely because so much of biomedical science is exploratory and experimental, scientific inquiry is not just thought but also action, action often involving research on living subjects, including human beings. And precisely because the use of technologies often has unintended or undesirable side effects, affecting many human goods in addition to health, safety, and the relief of suffering, large questions are necessarily raised when the goods promoted by technology come into conflict with others. For example, is the need to discover new cures for the sick a moral imperative that should trump all other goods and values? If not, then on what basis can it be limited? What moral boundaries should scientists and technologists respect as they continue their quests for knowledge and cures, whether or not they receive public funding? How can society establish and enforce such boundaries? And, on the other hand, how can science and technology be protected against unreasonable limitations imposed by excessively fearful legislators or overzealous regulators?

To be sure, these large questions are hard to answer in the abstract. As a result, they do not recommend themselves for much deliberation. Yet they are very close to the surface of the current debate about human cloning. Moreover, implicit answers to these questions, seldom articulated and rarely defended save by mere assertion, at least color and may even determine what people think should be done about human cloning. A clearer and more thoughtful awareness of the aims of biomedical science could help us assess whether and how human cloning might serve the ends of science and medicine and could help us more fully consider its possible benefits and potential drawbacks. But we must consider not only the ends of science, but also the means it employs. Cloning, after all, is a technique, a means of reaching some desired end. Even if the purposes it might serve are worthy, it must still be evaluated as a means. Not every means employed in the pursuit of worthy ends can pass ethical muster. This truth is widely recognized in the establishment of canons of ethics regarding the use of human subjects in research. It is also recognized in the established practice of technology assessment, which seeks to find the least problematic and least dangerous means for achieving a desirable end.

For instance, as a means of treating infertility or of providing a suitable source of compatible organs for transplantation, cloning raises difficulties having to do with human dignity and the costs of "manufacture" of the sort discussed earlier. Human cloning also raises questions about the ethics of research with human subjects, with risks of harm to the child-to-be, the egg donor, and the woman who would bring the cloned child to birth, questions that we shall take up in some detail in Chapter Five. Yet the most highly controverted moral argument about human cloning research involves a human subject not always considered when the ethics of research is discussed: the early human embryo. Because all cloning begins with the production of embryonic clones, and because such clones are potentially highly useful in biomedical research, questions of the ethics of means are absolutely central to the debate about the morality of cloning.

Ethical questions regarding the use of human embryos in research are, of course, not unique to cloning. They have been central to the recent and continuing controversy about federal funding of research on human embryonic stem cells, because human embryos produced by IVF offer possibilities for medical advances, beyond their use in assisted reproduction. The use of embryos has aided research on early human development. These embryos are also the source of human embryonic stem cells, pluripotent cells that may be induced to develop into all the tissues of the body. These stem cells thus may hold great promise for future treatment of chronic degenerative diseases and disabilities.

The difficulty arises because the embryos put to use in these ways are themselves destroyed. This fact raises serious and troubling questions about the proper way to regard these nascent human organisms and the morally appropriate way to treat them. Cloning techniques might provide an even more useful source of embryos for biomedical research than current IVF techniques. Human cloning could yield numerous identical embryos, could provide for the study of stem cells derived from individuals known to possess genetic diseases, and might eventually yield transplantable tissues for regenerative medicine that would escape immune rejection. Human cloning-for-biomedical-research therefore brings the moral question of means before us with even greater force. It calls on us to think of the good of medical advances and the relief of human suffering while at the same time considering our responsibilities to nascent human life and the possible harms to ourselves and future generations that may result from coming to regard the beginning stages of human life as raw material for use and exploitation.

While there is almost universal opposition to cloning-to-produce-children, the prospect of using cloned embryos in biomedical research has attracted significant support in the general public and among many scientists, patient advocacy groups, and policymakers. It therefore presents more complicated moral and policy challenges, and requires serious reflection on the duty of society to those of its members who are suffering, as well as its responsibility for nascent life. The precise character of both that duty and that responsibility is a subject of long-standing dispute, giving rise to a contentious but very important public debate.

CLONING AND PUBLIC POLICY

Beneath the current debate about human cloning lie major questions about the relation between science and technology and the larger society. Valuing freedom and innovation, our society allows scientists to inquire as they wish, to explore freely, and to develop techniques and technologies based on the knowledge they find, and on the whole we all benefit greatly as a result. We limit what scientists can do only in certain cases, as when their research requires the use of human subjects, in which case we erect rules and procedures to protect the health, safety, and dignity of the weak from possible encroachments by the strong. In more pervasive ways, we also shape what science does through public decisions about financial support and scientific education. With the uses of technology, we are sometimes more intrusive, establishing regulations to protect public health and safety or to preserve the environment. In rare cases, we even ban certain practices, such as the buying and selling of organs for transplantation. Yet, on the whole, the spirit of laissez-faire governs technological research, development, and use.

But when innovations arise that appear to challenge basic goods that we hold dear, or when the desirability of scientific and technological progress runs up against concerns for the protection of human life and well-being, we are forced to consider the tacit social contract between science and technology and the larger society. The current public and political deliberation about whether and how to restrict or prohibit human cloning forces us to do so in a most powerful way.

In addition, the current deliberation confronts us with the task of balancing important and commonly defended freedoms – the freedom of scientists to inquire, of technologists to invent, of individuals to reproduce, of entrepreneurs to invest and to profit – with the well-being of our society and its members. Circumstances in which otherwise beneficent freedoms can endanger paramount moral and social goods present serious challenges for free societies, and the prospect of cloning presents us with just such a challenge.

This is not an altogether unfamiliar challenge. There are other circumstances in which the freedom to explore, inquire, research, and develop technologies has been constrained. Biomedical science, as we have said, is restricted in its use of human subjects for research, and scientists are required to obtain informed consent and take great care to secure research subjects from harm. Scientific work is also restricted from activities that might harm the health of the general public, and from producing products that may endanger consumers. For example, the federal Food and Drug Administration sits at the juncture between development and marketing of medical products, regulating their introduction and use according to criteria of safety and efficacy. Our society has come to a near-total agreement on the need for such an agency and the importance of its work.

Human cloning, however, does not easily fall into any of the familiar classes of our experience with science. Nor do the ethical challenges it raises fit neatly into the categories of risks to health and safety that are ordinarily the basis of public oversight of science and technology. Raising ethical questions about ends as well as means, cloning is at once a potential human experiment, a possible aid to reproduction, an altogether new sort of procreative technique, a prospective means of human design, and a source of embryos and embryonic stem cells for research. It points back to familiar dilemmas of bioethics – including the ethics of human experimentation and embryo research – and it points forward to the sorts of challenges that will face us as biology gains greater technical prowess. It therefore invites us to think anew about the relationship between society and biomedical science and to evaluate the sufficiency of current institutions and practices that govern that relationship.

The potential dangers we face do not result from ill intent or bad faith. Neither of the prevailing caricatures in the cloning debate – the mad scientist on a blind quest for an inhuman immortality or the puritanical Luddite seeking to keep the future at bay – is accurate, appropriate, helpful, or fair. The challenge we face is not as easy as that. The challenge we face involves the conflict of competing sets of concerns and priorities, each in the service of vital human goods, and each driven by a desire to improve the human condition and to protect essential principles. The widely shared desire to cure disease, relieve suffering, understand human biology, and provide humankind with new and more powerful means of control can conflict, in this case, with the widely shared desire to respect life, individual identity, the dignity of human procreation, and other institutions and principles that keep our society healthy and strong. The challenge for our society is to determine, through public deliberation and thoughtful reflection, how best to adjudicate between these two desires and to determine what form to give to the tacit agreement between society and science, by which society promises freedom within bounds, and science affords us innovation, knowledge, and power while respecting reasonable limits.

The new and distinct challenges that confront us through cloning call upon us to consider the character of that tacit agreement, and to determine whether, and in what way, it might need to be amended and supplemented, especially in the face of the rapidly arriving new biomedical technologies that touch so directly upon our humanity. It is our hope in this report to contribute to just such a thoughtful consideration of the question.

9

Historical Aspects of Cloning

The previous chapter located human cloning in its larger human context. This chapter provides a brief history of human cloning, both as a scientific matter and as a subject of public discussion, debate, and legislation. Although we present only selected highlights, rather than a comprehensive account, we seek to enable the reader to place the present debate about cloning and this report into their proper historical setting. Until recently, all discussion of human cloning concentrated exclusively on the prospect of clonal reproduction, the production of individuals genetically virtually identical to previously existing ones. Our historical account here reflects that emphasis. Yet we will also consider the emerging interest in cloning-for-biomedical-research, a prospect connected to the recent isolation of embryonic stem cells and their potential for the understanding and treatment of human disease and disability.

SCIENTIFIC MILESTONES

As a scientific and technical possibility, human cloning has emerged as an outgrowth of discoveries or innovations in developmental biology, genetics, assisted reproductive technologies, animal breeding, and, most recently, research on embryonic stem cells. Assisted reproductive techniques in humans accomplished the in vitro fertilization of a human egg, yielding a zygote and developing embryo that could be successfully implanted into a woman's uterus to give rise to a live-born child. Animal breeders developed and refined these techniques with a view to perpetuating particularly valuable animals and maintaining laboriously identified genomes. Most recently, the isolation of embryonic stem cells and their subsequent in vitro differentiation into many different cell types have opened up possibilities for repairing and replacing diseased or nonfunctioning tissue, and thus possible research uses for cloned human embryos.

The German embryologist Hans Spemann conducted what many consider to be the earliest "cloning" experiments on animals. Spemann was interested in answering a fundamental question of biological development: does each differentiated cell retain the full complement of genetic information present initially in the zygote? In the late 1920s, he tied off part of a cell containing the nucleus from a salamander embryo at the sixteen-cell stage and allowed the single cell to divide, showing that the nucleus of that early embryo could, in effect, "start over." In a 1938 book, *Embryonic Development and Induction*, Spemann wondered whether more completely differentiated cells had the same capacity and speculated about the possibility of transferring the nucleus from a differentiated cell – taken from either a later-stage embryo or an adult organism – into an enucleated egg. As he explained it: "Decisive information about this question may perhaps be afforded by an experiment which appears, at first sight, to be somewhat fantastical. This experiment might possibly show that even nuclei of differentiated cells can initiate normal development in the egg protoplasms." But Spemann did not know how to conduct such an experiment.

Research with frogs fourteen years later encouraged progress toward the "fantastical experiment." In 1952, the American embryologists Robert Briggs and Thomas J. King first successfully transferred nuclei from early embryonic cells of leopard frogs to enucleated leopard frog eggs. The "activated egg" began to divide and develop, became a multicellular embryo, and then became a tadpole. Embryologists in other laboratories successfully repeated these initial experiments on different species of frogs. But additional experience also showed that the older and more differentiated a donor cell becomes, the less likely it is that its nucleus would be able to direct development.

In 1962, the British developmental biologist John Gurdon reported that he had produced sexually mature frogs by transferring nuclei from intestinal cells of tadpoles into enucleated frog eggs. The experiments had a low success rate and remained controversial. Gurdon continued this work in the 1970s, and he was able to produce tadpoles by transferring the nucleus of adult frog skin cells into enucleated frog eggs. Later experiments established that many factors in addition to the intact nucleus are crucial to success (see Chapter Four for further discussion). In retrospect, it is surprising that any of these earlier experiments produced positive results. But despite their low success rates, these experiments demonstrated that the nucleus retained its full complement of genetic information and encouraged later investigators to explore mammalian cloning.

The birth of Louise Brown in 1978, the first baby conceived through in vitro fertilization (IVF), was also an important milestone, because it demonstrated that human birth was possible from eggs that were fertilized outside the body and then implanted into the womb. As for the possibility of cloning animals from adult cells – especially mammals – the work in the intervening years focused largely on the reprogramming of gene expression in somatic cells, the transfer of nuclei taken from embryos in mammals (beginning with mice in the 1980s), and finally the work of Ian Wilmut and his colleagues at the Roslin Institute with adult nuclei, which led to the birth of Dolly on July 5, 1996. Since then, similar success has been achieved in cloning other mammalian species, including cattle, goats, pigs, mice, cats, and rabbits (see Chapter Four).

The animal cloners did not set out to develop techniques for cloning humans. Wilmut's goal was to replicate or perpetuate animals carrying a valuable genome (for example, sheep that had been genetically modified to produce medically valuable proteins in their milk). Others, such as the cloners of the kitten CC, were interested in commercial ventures for the cloning of pets. Yet the techniques developed in animals have encouraged a small number of infertility therapists to contemplate and explore efforts to clone human children.

And, following the announcement in 1998 by James Thomson and his associates of their isolation of human embryonic stem cells, there emerged an interest in cloned human embryos, not for reproductive uses but as a powerful tool for research into the nature and treatment of human disease.

HUMAN CLONING FROM POPULAR LITERATURE TO PUBLIC POLICY: FROM BRAVE NEW WORLD TO THE BIRTH OF DOLLY

Technological novelties are often imagined and discussed in literature, especially in science fiction, before they are likely or even possible in practice. This has certainly been the case with human cloning, whose place in the popular imagination precedes the earliest successful animal cloning experiments. Perhaps the most famous early modern account of human cloning is Aldous Huxley's Brave New World (1932), where natural human procreation has become a thing of the past, and where babies are produced in identical batches through "Bokanovsky's Process." As the novelist tells it:

One egg, one embryo, one adult – normality. But a bokanovskified egg will bud, will proliferate, will divide... becoming anywhere from eight to ninety-six embryos – a prodigious improvement, you will agree, on nature. Identical twins – but not in piddling twos and threes... Standard men and women; in uniform batches.

The relevance or irrelevance of Huxley's vision to the dilemmas of the present is of course a matter of serious disagreement. Some believe that fears of a "Brave New World" are fantasy divorced from both the political realities of modern liberal democracy and the facts of science. Others believe the book remains a prescient warning of where biological self-manipulation could take us – which is to say, to a world where family is obsolete, life is engineered to order in the laboratory, and human beings have reduced themselves to well-satisfied human animals.

In the late 1960s, following John Gurdon's successful cloning experiments, a more focused debate on both the likelihood and the ethical and social implications of human cloning began among scientists, theologians, and ethicists. At this time, the still hypothetical possibility of cloning humans was considered as a part of a broader eugenic project to improve the genetic stock of humans as a species. In a famous article published in *The American Naturalist* in 1966, entitled "Experimental Genetics and Human Evolution," Nobel laureate biologist Joshua Lederberg described what he took to be the prospects of "clonal reproduction." "Experimentally," he wrote, "we know of successful nuclear transplantation from diploid somatic as well as germline cells into enucleated amphibian eggs. There is nothing to suggest any particular difficulty about accomplishing this in mammals or man, though it will rightly be admired as a technical tour-de-force when it is first implemented." He also predicted "there will be little delay between demonstration and use."

While Lederberg concluded his essay by exhorting his readers not to "mistake comment for advocacy," he clearly believed that clonal reproduction might offer a number of human benefits or improvements. "If a superior individual (and presumably then genotype) is identified, why not copy it directly, rather than suffer all the risks of recombinational disruption, including those of sex," he asked. "The same solace is accorded the carrier of genetic disease: why not be sure of an exact copy of yourself rather than risk a homozygous segregant; or at worst copy your spouse and allow some degree of biological parenthood." He described other possibilities – such as "the free exchange of organ transplants with no concern for graft rejection" and more efficient communication between individuals in "stressed occupations."

In the end, Lederberg argued that "tempered clonality" – a mix of clonal and sexual reproduction – might, at least from a biological standpoint, "allow the best of both worlds – we would at least enjoy being able to observe the experiment of discovering whether a second Einstein would outdo the first one." Nevertheless, he acknowledged the possibility for "social frictions" and ethical dilemmas that might result from clonal reproduction – including whether "anyone could conscientiously risk the crucial experiment, the first attempt to clone a man." He suggested that the "mingling of individual human chromosomes with other mammals assures a gradualistic enlargement of the field and lowers the threshold of optimism or arrogance, particularly if cloning in other mammals gives incompletely predictable results." And he feared that social policy might become based on "the accidents of the first advertised examples" rather than "well-debated principles."

In 1970, the theologian and ethicist Paul Ramsey responded to Lederberg's portrait of human cloning – and, more generally, to the prospects for human self-modification – in a book called *Fabricated Man: The Ethics of Genetic Control*. He argued that human cloning would violate the ethical responsibilities of both science and parenthood: it would involve experiments on the child-to-be; it would transform parenthood into manufacture; and it would burden children with the genetic predisposition of their "maker" and so deny the cloned child a unique independence in the very act of bringing him or her to life. "[T]o attempt to soar so high above an eminently human parenthood," Ramsey wrote, "is inevitably to fall far below – into a vast technological alienation of man .... The entire rationalization of procreation – its replacement by replication – can only mean the abolition of man's embodied personhood."

Ramsey believed that such a willingness to experiment on human life – or to create sub-humans-showed how the effort to perfect and improve humankind through genetic control leads in fact to ethical coarsening and to a disregard for actual human beings. "In the present age," he wrote, "the attempt will be made to deprive us of our wits by comparing objections to schemes of progressive genetic engineering or cloning men to earlier opposition to inoculations, blood transfusions, or the control of malaria. These things are by no means to be compared: the practice of medicine in the service of life is one thing; man's unlimited self-modification of the genetic conditions of life would be quite another matter."

The debate over human cloning and genetic manipulation continued in the early 1970s. The Nobel laureate geneticist James D. Watson testified before Congress in 1971 on the subject of human cloning. He described the science that was taking us there, including John Gurdon's success in cloning frogs and the work of R. G. Edwards and P. S. Steptoe "in working out the conditions for routine test-tube conception of human eggs." "Human embryological development," Watson observed, "need no longer be a process shrouded in secrecy. It can become instead an event wide-open to a variety of experimental manipulations." Watson called for the creation of national and international committees to promote "wide-ranging discussion ... at the informal as well as formal legislative level, about the manifold problems which are bound to arise if test-tube conception becomes a common occurrence." "This is a decision not for the scientists at all," he said. "It is a decision of the general public – do you want this or not?" and something that "if we do not think about it now, the possibility of our having a free choice will one day suddenly be gone."

In 1972, Willard Gaylin, a psychiatrist and co-founder of the newly formed Institute of Society, Ethics, and the Life Sciences (later called the Hastings Center), made James Watson's warnings about cloning even more dramatic – with a *New York Times Magazine* article titled "The Frankenstein Myth Becomes a Reality – We Have the Awful Knowledge to Make Exact Copies of Human Beings." Gaylin hoped that the prospect of human cloning would awaken the public – and the scientific community – to the larger ethical implications of the life sciences. The same year, biologist and ethicist Leon R. Kass published an essay in *The Public Interest* called "Making Babies – The New Biology and the 'Old' Morality," which addressed the prospect of both in vitro fertilization and human cloning, and wondered whether "by tampering with and confounding [our] origins, we are involved in nothing less than creating a new conception of what it means to be human."

In stark contrast to Gaylin and Kass, ethicist Joseph Fletcher argued that human cloning would not be dehumanizing at all, but would, in a number of circumstances, serve the good of both society and individuals. In his 1974 book *The Ethics of Genetic Control: Ending Reproductive Roulette*, he argued that "Good reasons in general for cloning are that it avoids genetic diseases, bypasses sterility, predetermines an individual's gender, and preserves family likenesses. It wastes time to argue over whether we should do it or not; the real moral question is when and why." For Fletcher – unlike Ramsey, Gaylin, and Kass – genetic control would serve the human end of self-mastery and self-improvement, it would improve the quality of life for individuals, and it would aid the progress of the human species. Gunther Stent, a molecular biologist at the University of California at Berkeley, echoed this view that human cloning would contribute to human perfection. As he wrote in a 1974 article in Nature: "To oppose human cloning... is to betray the Western dream of the City of God. All utopian visionaries, from Thomas More to Karl Marx, think of their perfect societies as being populated not by men but by angels that embody all of the best and none of the worst human attributes." With cloning, he suggested, such a city might one day be possible.

For several years, cloning remained a topic for fiction and philosophy, but fantasy had yet to turn into fact. In 1978, in a book titled *In His Image: The Cloning of a Man,* science writer David Rorvik claimed that he was involved in a secret project to clone a millionaire in Montana named "Max." The book caused a flurry of reaction – ranging from horror to amusement to nearly universal skepticism and denunciation in the scientific community – and eventually led to hearings before Congress on May 31, 1978. Robert Briggs, who with Thomas King cloned the first frog embryo from blastula frog cells in 1952, declared that the work in frogs demonstrated not that human cloning is now or imminently possible, but that "cloning in man or any other animal is not just a technical problem to be solved soon but may, in fact, never occur." James Watson, who just a few years earlier had urged a national conversation and possible legislation on human cloning because of the rapid advances in the science, declared that we would "certainly not [see the cloning of a man] in any of our lifetimes. I wouldn't be able to predict when we might see the cloning of a mouse, much less a man." Rorvik eventually admitted that the book was a hoax.

In the years that followed, claims and counter-claims of scientific advances in mammalian cloning – including the controversy beginning in 1981 over whether any of several independent laboratories had actually cloned mice – prompted more public reaction and discussion about the issue. But there was no sustained or widespread public interest, and cloning lost its prominent place within the bioethics literature. The President's Bioethics Commission, in its 1982 report *Splicing Life,* briefly discussed human cloning as well as IVF, but held that both were beyond the scope of that report because they could be considered reproductive technologies that did not necessarily involve modifying the genome (pp. 9-10). With regard to human cloning in particular, the report added that the possibility had received a good deal of public attention and it was therefore important to emphasize that even if it ever did become possible in humans, it would not result in an identical being.

The National Institutes of Health Human Embryo Research Panel, which issued a report in 1994 on federal funding for research involving preimplantation human embryos, deemed research involving nuclear transplantation, without transfer of the resulting cloned embryo to a uterus, as one type of research that was acceptable for federal support. The report noted that the majority on this point was narrow, with nearly as many panel members concluding that the ethical implications of nuclear transplantation should be studied further before any such research could be acceptable for federal funding (Exec. Summ., p. xvii). In its discussion of cloning techniques, the panel noted that many different procedures are all called "cloning," and said in a footnote, "Popular notions of cloning derive from science fiction books and films that have more to do with cultural fantasies than actual scientific experiments."

Of course, there had been, in the meantime, continued scientific work in nuclear transplantation in animals – including mammals. And with the 1997 announcement of the cloning of Dolly, the prospect of human cloning once again became a prominent issue in public discussion, debate, and public life.

THE HUMAN CLONING DEBATE: FROM DOLLY TO THE PRESENT

In late February 1997, Ian Wilmut and his team at the Roslin Institute in Scotland announced that they had, by means of somatic cell nuclear transfer, successfully cloned the first mammal from an adult somatic cell – Dolly the sheep. President Bill Clinton and British Prime Minister Tony Blair immediately denounced any attempts to clone a human being, and the President asked his National Bioethics Advisory Commission (NBAC) to report within ninety days on the scientific, ethical, and legal questions surrounding the prospect of human cloning. Congress likewise held a series of hearings – the first one on March 12, 1997. A widespread – though not universal – consensus emerged that attempts to clone a human being would at present be irresponsible and immoral. As Wilmut explained before Congress, "Our own experiments to clone sheep from adult mammary cells required us to produce 277 'reconstructed' embryos. Of these, twenty-nine were implanted into recipient ewes, and only one developed into a live lamb. In previous work with cells from embryos, three out of five lambs died soon after birth and showed developmental abnormalities. Similar experiments with humans would be totally unacceptable."

Most ethicists agreed, though for different reasons. All agreed that cloning attempts on human beings "at this time" would be reckless experiments on the child-to-be and therefore totally unjustified. Many stressed, as Ramsey, Gaylin, and Kass had done in the 1970s, that human cloning would undermine the human meaning of parenthood and identity; that it would mean a giant step toward genetic engineering, creating the first children whose genetic predisposition was known and selected in advance; and that it would turn procreation increasingly into a form of manufacture.

In contrast, some bioethicists, including John Robertson and Ruth Macklin, believed that human cloning presented no inherent threat to public or private morality, that government had no legal authority or justification for banning clonal reproduction, and that it must be judged in terms of its particular uses, not dismissed outright.

In June 1997, NBAC released its report *Cloning Human Beings*, which concluded that

At present, the use of this technique to create a child would be a premature experiment that would expose the fetus and the developing child to unacceptable risks. This in itself might be sufficient to justify a prohibition on cloning human beings at this time, even if such efforts were to be characterized as the exercise of a fundamental right to attempt to procreate.

NBAC also pointed to other moral concerns "beyond the issue of the safety of the procedure," including "the potential psychological harms to children and effects on the moral, religious, and cultural values of society" that "merit further discussion." NBAC recommended a three-to-five-year federal moratorium on human cloning – stating that the consensus came from the fact that the technique was not yet safe – to be revisited and reevaluated after that time. "Whether upon such further deliberation our nation will conclude that the use of cloning techniques to create children should be allowed or permanently banned is, for the moment, an open question."

In early 1998, the United States Senate considered legislation, proposed by Republican Senators Christopher Bond of Missouri, Bill Frist of Tennessee, and Judd Gregg of New Hampshire, to ban all human cloning permanently. Nearly all senators denounced clonal reproduction, but many believed that the proposed ban, which would have made it illegal to create human embryos by means of somatic cell nuclear transfer, would undermine potentially valuable scientific research. Democratic Senators Edward Kennedy of Massachusetts and Tom Harkin of Iowa led the opposition, with the widespread support of patient advocacy groups, scientific and medical organizations, and the biotechnology industry. As Senator Kennedy put it:

> Every scientist in America understands the threat this legislation poses to critical medical research. Every American should understand it, too.... Congress can and should act to ban cloning of human beings during this session. But it should not act in haste, and it should not pass legislation that goes far beyond what the American people want or what the scientific and medical community understands is necessary or appropriate.

The legislation died after heated debate, and the concern over human cloning temporarily lost urgency and subsided.

In November 1998, a new scientific discovery was unveiled that would soon provoke a different public policy debate, one that would become entangled with the ethical and social questions surrounding human cloning. James Thomson and John Gearhart separately announced the isolation of human embryonic stem cells – multipotent cells (see Glossary of Terms) derived from human embryos that they believed hold great promise for curing or treating many diseases and injuries. The discovery led to another wave of hearings on, and interest in, the ethics of biological science. It also renewed debate over whether embryo research should be eligible for public funding (since 1996, Congress had prohibited federal funding of research involving the destruction of human embryos). One subject under consideration was the possible future use of cloned human embryos for stem cell research, which some scientists believed might be uniquely useful for understanding embryological development and genetic disease and for possible use in stem cell therapies.

In August 2000 – after another NBAC study – President Clinton announced new guidelines that would have altered the ban on federal funding of embryo research. The new guidelines, proposed by the National Institutes of Health, stipulated that the agency would fund research on embryonic stem cells so long as public funds were not used to destroy the embryos, the embryos were left over from IVF clinics, and donors of the embryos consented to the research.

In early 2001, President George W. Bush announced that he would review these guidelines rather than implement them immediately. Around the same time, a number of pro-cloning groups and fertility doctors – including the Raelians, who believe that humans are the products of cloning by aliens – announced their intention to clone human beings by the end of the year.

Other individuals and scientific organizations worked to protect possible cloning research from future restrictions, though some scientists (such as Rudolf Jaenisch and Ian Wilmut) publicly argued against cloning-to-produce-children. A flurry of hearings on human cloning soon followed – the first one in the House of Representatives on March 28, 2001, and continuing in both the House and the Senate throughout the summer. The hearings addressed cloning-to-produce-children as well as issues related to cloning-for-biomedical-research.

Two general approaches to banning human cloning emerged. The first approach, proposed in a bill sponsored by Republican Representative David Weldon of Florida and Democratic Representative Bart Stupak of Michigan in the House, and Republican Senator Sam Brownback of Kansas and Democratic Senator Mary Landrieu of Louisiana in the Senate, called for a ban on all human cloning, including the creation of cloned embryos for biomedical research.

The second approach, proposed in a bill sponsored by Republican Senators Arlen Specter of Pennsylvania and Orrin Hatch of Utah and Democratic Senators Diane Feinstein of

California and Edward Kennedy of Massachusetts, sought to prohibit human reproductive cloning, while allowing the use of cloning technology to produce stem cells, by making it illegal to implant or attempt to implant cloned human embryos "into a uterus or the functional equivalent of a uterus."

On July 31, 2001, the House of Representatives passed the Weldon – Stupak bill (the ban on all human cloning) by a vote of 265 to 162. In November 2001, scientists at Advanced Cell Technology, Inc., of Worcester, Massachusetts, one of the leading commercial advocates of cloning-for-biomedical-research, reported what they claimed were the first cloned human embryos. The announcement – along with continued debate on the possible use of cloned human embryos for stem cell research – left the issue in the United States Senate, where it stands as of this writing.

Meanwhile, the general public has consistently expressed the view that human cloning is wrong – most recently, a Gallup poll from May 2002 that showed opposition to cloning to produce a child at 90 percent, and opposition to "cloning of human embryos for use in medical research" at 61 percent. Asked about medical research using stem cells obtained from human embryos (with no mention of how the embryo was generated), 52 percent found it morally acceptable, while 51 percent found acceptable the "cloning of human cells from adults for use in medical research."

In addition to activity at the federal level, many states have been active. As of this writing, twenty-two states have considered various policy alternatives on cloning, and six have passed legislation.

Several nations, including Denmark, France, Norway, Spain, and Canada have passed or sought either partial or total bans. For example, in the United Kingdom, cloning-to-produce-children is forbidden but cloned embryos up to fourteen days old may be used in biomedical research. In Germany, all human cloning is forbidden by law. There are also efforts now at the United Nations and other international organizations to pass a world-wide ban on human cloning – with many of the same disagreements internationally as there are nationally about what kind of ban to pass.

ON TERMINOLOGY

The rest of this chapter will develop the meaning of these terms and provide the analysis and argumentation that have led us to these choices. Because there is much to be learned about the subject through the discussion of alternative terminologies, and because we believe strongly that the judicious use of language is necessary for sound moral choice, we present our discussion of this matter at some length.

Fruitful discussion of the ethical and policy issues raised by the prospects of human cloning – as with any other matter – can proceed only if we can find appropriate and agreed-upon terms for describing the processes and products involved. Before we can get to possible moral or policy arguments or disagreements, we need to agree about what to call that about which we are arguing. As a contribution to public understanding, we emphasize that this is not an easy thing to do, and we indicate how and why we have gone about making our terminological choices.

What exactly is meant by the term "cloning"? What criterion justifies naming an entity a "clone"? How is the term "cloning" related to what scientists call "somatic cell nuclear transfer (SCNT)" or "nuclear transplantation"? What should we call the single-cell entity that results from SCNT, and what should we call it once it starts to divide and develop? How, if at all, should our names for such activities or such entities be affected by the purposes we have for engaging in the activities or for using the entities?

As these questions imply, there is much confusion today about the terms used in discussing human cloning. There is honest disagreement about what names should be used, and there are also attempts to select and use terms in order to gain advantage for a particular moral or policy position. One difficulty is the difference between the perspective of science and the perspective of lived human experience. People who look at the phenomena of human reproduction and development through the lens of science will see and describe things in terms that often differ widely from those in ordinary usage; moreover, when an ordinary term is used in scientific parlance, it sometimes is given a different meaning. Similar divergences are possible also for people who look at these matters through the lens of different cultural, philosophical, or religious beliefs. Yet at the same time, all of us–scientists or not, believers or not – encounter these same matters on the plane of lived human experience, for which the terms of everyday speech may well be more suitable. Because this same common (nonscientific) discourse is also the medium of discourse for the ethical and policy discussions, we shall strive to stay close to common speech, while at the same time making the best use we can of scientific findings to avoid mistakes and misconceptions.

Advisers to decision makers should strive not only for accuracy, but also for fairness, especially because the choice of names can decisively affect the way questions are posed and, hence, how answers are given. The issue is not a matter of semantics; it is a matter of trying fairly to call things by names that correctly describe them, of trying to fit speech to fact as best one can. For the sake of clarity, we should at least stipulate clearly the meanings we intend by our use of terms. But we should also try to choose terms that most accurately convey the descriptive reality of the matter at hand. If this is well done, the moral arguments can then proceed on the merits, without distortion by linguistic sloppiness or chicanery.

Many of the terms that appear in the debate about cloning are confusing or are used in a confused manner.

First, there are difficulties concerning the terms that seek to name the *activity* or *activities* involved: cloning, asexual reproduction, reproductive cloning, nonreproductive cloning, research cloning, therapeutic cloning, somatic cell nuclear transfer (or nuclear transplantation), nuclear transfer for stem cell research, nuclear transplantation to produce stem cells, nuclear transfer for regenerative medicine. At stake are such questions as whether all acts of SCNT should be called cloning. Some worry that the term "cloning" unfairly prejudices people against the activity when it is used to describe research activities.

Second, there are difficulties concerning the terms that seek to name the *entity* or *entities* that result from human cloning (or human SCNT): cell, egg, activated cell, totipotent cell, clonote, reconstituted (or reconstructed) egg, zygote, clump of cells, embryo, human embryo, human organism, blastocyst, clonocyst, potential human being, human being, human clone, person. At stake here is the nature – and the possible *moral* status – of the entities that are involved in the subsequent manipulations, whether for producing a child or for use in biomedical research. Some worry that use of any term but "embryo" will unfairly prejudice people in favor of embryo-destructive activities by hiding from view the full import of the activity.

Third, there are difficulties concerning the terms that seek to describe the *relation* between the cloned entity and the person whose somatic cell nucleus was transferred to produce the cloned entity: genetic copy, replica, genetically virtually identical, noncontemporary twin, delayed genetic twin, clone.

TOOLS OF ANALYSIS

As a prelude to examining the *activity* or the *deed* of cloning, some general analytical observations will be helpful. Although all aspects of an activity or action are relevant to understanding its full human meaning, when describing a deed it is sometimes useful to distinguish *what* it is from both *how* it is done and *why* it is done. The act itself (*what*) may be accomplished by a variety of means or techniques (*how*), and it may be undertaken for a variety of motives or purposes (*why*). To be sure, there is a danger of distortion in this disaggregating analysis of human activity, and there is disagreement about the degree to which the motives or purposes of the agent are to be reckoned in the description of the act itself. People argue, for example, whether "mercy killing" differs *as an act* from murdering a rival (or executing a murderer or killing someone in self-defense), or whether they are all equally acts of homicide (literally, "killing a human being") whose *moral* meaning ("Is it justified or not?" "Is it wrong or not?") we can then proceed to debate, if we wish, by attending not only to the bare act of taking a human life but also to the agent's motive and purpose. Though we do not wish to beg this question, the very existence of this disagreement suggests that we do well not to ignore the naked act itself, for it may have a meaning independent of what moved the agent, a meaning relevant to subsequent moral assessment that we do not wish to overlook.

To illustrate: in vitro fertilization (IVF: the merging of egg and sperm outside the human body [*in vitro* = "in glass"], yielding a zygote that is the beginning stage of a new living being) is the deed (*what*). It is an act of "fertilization," of making fertile, of making the egg cell ready and able to develop into a human organism. This fertilization may be accomplished in at least two ways (*how*): by merely mixing egg and sperm, allowing the sperm to find and penetrate the egg, or by the technique of injecting individual sperm directly into the egg (a technique known as intracytoplasmic sperm injection, ICSI). And it may be done for the (proximate) purpose (*why*) of initiating a pregnancy, in turn for the (ultimate) purpose of providing a child for an infertile couple; or it may be done for the (proximate) purpose of providing living human embryos for basic research on normal and abnormal embryological development, in turn for the (ultimate) purposes of understanding human development or of discovering cures for diseases and producing tissues for regenerative medicine. Though the technique used or the purposes served may differ, in *one* crucial respect the deed (IVF) remains the same and bears a common *intrinsic meaning*: a human zygote, the first stage of a new human being, is intentionally produced outside the body and exists in human hands and subject to human manipulation.

As it happens, this fact is more or less accurately reflected in the descriptive terminology used for IVF. Interestingly enough, unlike the situation with cloning, no one distinguishes between "reproductive IVF" and "therapeutic IVF" or "research IVF," naming the activity or deed after the motive or purpose of the agent. This may reflect the historical fact that IVF was initiated by people who were interested in using it to produce live-born children for infertile couples; the research use of "surplus" embryos produced by IVF came only later. But it happens that this common name is also descriptively apt and remains so regardless of why IVF was done in a particular case: the deed *is* fertilization of egg by sperm, producing a living human zygote, the first stage of the development of a new human being.

It should be noted that, although we began by trying to describe the deed rather than the product of the deed, the two aspects merged necessarily. The meaning of the act of "fertilization" falls forward onto the nature of the "object" that fertilization produces: the fertilized egg or zygote or earliest embryo. (By contrast, there is nothing in the name of the technique "intracytoplasmic sperm injection" that even hints at the immediate result or goal of the intended injection.) Similar attention to the nature of the product may turn out to be indispensable for a proper characterization of the activity of cloning.

CLONING: TOWARD AN APPROPRIATE TERMINOLOGY

Though much of the terminological confusion and controversy concerns the way to describe the different *kinds* of cloning practices that are envisioned, the term "cloning" itself is not without its own ambiguities. A "clone" (noun, from the Greek *klon*, "twig") refers to a *group* of genetically identical molecules, cells, or organisms descended from a single common ancestor, as well as to *any one* of the one or more individual organisms that have descended asexually from a common ancestor. Both the group and each of its members are "a clone." "To clone" (verb) is to duplicate or produce a genetic duplicate or duplicates of a molecule, cell, or individual organism.

The replication of DNA fragments in the laboratory is called "DNA cloning." The physical isolation of a single cell and its subsequent multiplication in tissue culture into a population of descendants is referred to as "single cell cloning." The laboratory culture of bacteria and the asexual propagation of plants by means of cuttings are instances of organismal cloning. Cloning of higher organisms is more complex: all cloning of vertebrate organisms must begin at the embryonic stages. Contrary to what some people imagine, cloning of amphibians or mammals (including human beings) is not the direct duplication ("photocopying") of an adult organism.

In the sense relevant here, "cloning" is a form of asexual reproduction (parthenogenesis is another), the production of a new individual not by the chance union of egg and sperm but by some form of replication of the genetic makeup of a single existing or previously existing individual. (In biological or functional terms, the core of sexual reproduction is not bodily intercourse but the fusion of male and female *germ cells*; thus IVF, though it takes place outside the body, is – biologically speaking – a form of sexual reproduction.) Cloning is the activity of producing a clone, an individual or group of individuals genetically virtually identical to the precursor that is being "replicated."

CLONING-TO-PRODUCE-CHILDREN; CLONING-FOR-BIOMEDICAL-RESEARCH

In much of the current public discussion, we encounter a distinction between two sorts of cloning: "reproductive" and "therapeutic." The distinction is based entirely on the differing goals of the cloners: in the first case, the goal is the production of a (cloned) child; in the second case, the development of treatments for diseases (suffered not by the clone, but by others). We recognize the distinction and the need for terms to describe the difference. But the terms currently in vogue have their difficulties. Both terms have been criticized by partisans of several sides of the debate, and for understandable reasons.

Some object to the term "reproductive cloning" used as a term of distinction, because they argue that *all* cloning is reproductive. Their reason: all human cloning intends and issues in the production of a cloned human embryo, a being distinct from the components used to generate it, a new human being in the earliest stage of development or "reproduction." (This claim, we would suggest, is at this stage a descriptive point, not yet a normative one; *it does not necessarily imply that such a being is fully human or "one of us," hence deserving of the moral and social protection accorded "persons."*) The fact that only some of these embryonic cloned humans are wanted for baby-producing purposes does not, in the view of these critics, alter this fact about their being. In support of their claim that cloning occurs (only) at the beginning, they note that once the cloning act of nuclear transfer has occurred, all new influences that act upon the new human organism cease to be "genetic" (nature) and are now "environmental" (nurture). Instead of "reproductive cloning," we shall speak of "cloning-to-produce-children."

Others object to the term "therapeutic cloning" for related reasons. The act of cloning embryos may be undertaken with healing motives. But it is not *itself* an act of healing or therapy. The beneficiaries of any such acts of cloning are, at the moment, hypothetical and in the future. And if medical treatments do eventually result, the embryonic clone from which the treatment was derived will not itself be the beneficiary of any therapy. On the contrary, this sort of cloning actually takes apart (or destroys) the embryonic being that results from the act of cloning.

To avoid the misleading implications of calling any cloning "therapeutic," we prefer the terms "research cloning" or "cloning for research," which also more accurately indicate the purpose of the activity. Yet some may find fault with this replacement.

Because it appears to be a deliberate substitution for "therapeutic cloning," it may seem to imply that the scientists have abandoned the pursuit of medical cure in favor of research as an end in itself. Believing that producing cloned embryos just for research would seem to be less justifiable than producing them with healing motives, these critics of the term "research cloning" want to avoid giving the impression that scientists want to experiment on new life just to satisfy their curiosity. We believe that this legitimate concern can be addressed by appending the adjective "biomedical" to make clear that the aim of the research is to seek cures and treatments for human diseases. We therefore opt to use the term "cloning-for-biomedical-research."

Some proponents of the activity called "therapeutic cloning" also now object to the term, but not because of the adjective. Though it was proponents who originally coined and used the term, some of them now want to shed the term "cloning," fearing that the bad or distressing connotations of the latter will weigh against the activity itself. Cloning, they insist, should be reserved for the activity that produces live-born cloned babies; it should not apply to the initial act that starts the process, which they would rather call "somatic cell nuclear transfer" or "nuclear transplantation."

The reason for such redescription is not wholly cosmetic and rhetorical; because the researchers are primarily interested in obtaining pluripotent stem cells, their focus is on the somatic cell nucleus and what must be done to it (transfer or transplantation) in order for it to revert to the undifferentiated condition of the primordial stem cell stage. Nevertheless, such terminological substitution is problematic, for the following reasons.

Although as a scientific matter "somatic cell nuclear transfer" or "nuclear transplantation" may accurately describe the *technique* that is used to produce the embryonic clone, these terms fail to convey the nature of the deed itself, and they hide its human significance. The deed, fully described, is the production of a living human entity (or "embryo" or "organism"; of the right name for the product, more later) that is genetically virtually identical to the donor organism, a fact or meaning not captured in the name for the technique or method, the transfer of a somatic cell nucleus (into an unfertilized egg whose own nucleus has been removed or inactivated). As a name, SCNT is not a fully accurate description even of the technique itself. It makes no reference to the intended and direct result of the deed of nuclear transfer. It also omits mention of the fact that the recipient of the transferred nucleus is an (enucleated) *egg* cell (rather than another kind of cell), which then can be made to initiate cell division as if it were just like a zygote produced by fertilization. The further amendments, "somatic cell nuclear transfer *for stem cell research*" or "nuclear transplantation *for regenerative medicine*" or "nuclear transplantation *to produce stem cells*" only compound the difficulty, mixing in the purpose of the activity with its technique, thus further obscuring the immediate meaning of the act itself, the production of a living cloned human embryo.

Cloned Human Embryo: The Product of SCNT

What shall we call the product of SCNT? The technical description of the cloning method (that is, SCNT) omits all reference not only to cloning but also to the immediate product of the activity. This obscurity enables some to argue that the *immediate product* of SCNT is not an "embryo" but rather "an egg" or "an unfertilized egg" or "an activated cell," and that the subsequent stages of development should not be called embryos but "clumps of cells" or "activated cells." To be sure, there are genuine difficulties and perplexities regarding what names to use, for we are dealing with an entity new in our experience. Partly for this reason, some people recommend avoiding the effort to describe the nature of the product, preferring instead to allow the *uses we human beings have for it* to define its being, and hence its worth. But, for reasons of both truth and ethical conduct, we reject this approach as improper. We are all too familiar with instances in which some human beings have defined downward the status of other beings precisely to exploit them with impunity and with a clear conscience. Thus, despite the acknowledged difficulties in coming to know it accurately, we insist on making the effort to describe the product of SCNT as accurately and as fairly as we can.

The initial product of SCNT is a single cell, but it is no ordinary cell. It is also an "egg" and a "reconstituted egg." But even that is not the whole story. The "reconstituted" egg is *more* than reconstituted; it has been capacitated for development. Because the egg now has a diploid nucleus, it has become something beyond what it was before: it now contains in a single nucleus the full complement of genetic material necessary for producing a new organism. And being an *egg* cell, it uniquely offers the cytoplasmic environment that can support this development. The product of SCNT thus resembles and can be made to act like a fertilized egg, a cell that not only has the full complement of chromosomes but also is capable (in animals) or may be capable (in humans) of developing into a new organism. In other words, in terms of its future prospects, it is a "zygote-like entity" or a (cloned) "zygote equivalent."

The initial product of SCNT is, to be sure, not just a cell but an *active* cell. (More precisely, it is a cell that can be activated by electric stimulation.) But "activated cell" is much too vague to describe the activity of which it is capable. For, once stimulated, the activity of this "cell" produced by SCNT is nothing other than human embryological development, initiated and directed by the cell itself. The processes of cellular growth, chromosomal replication, cell division, and (ultimately) differentiation into the tissues and organs of the organism are coordinated processes under the governance of the immanent developmental plan encoded in the cell's genetic material. In other words, the product of SCNT is an organism in its germinal stage, and its activities are those of an integrated and self-developing whole.

Another suggested name, better than "activated cell," is "totipotent cell" – a cell that is "capable of all." But this too is ambiguous. If what is meant is that it can (and will, should it be stimulated to do so) become "any and all" of the different kinds of cells in the body, then it is an insufficient meaning. For, as explained in the previous paragraph, *this* totipotent cell may also become the "all" that is the integrated *whole* (cloned) mature organism itself (along with a portion of the placenta that would give it nourishment). In this second and fuller meaning of "totipotent," a totipotent cell is then just a *functional* synonym for the "zygote": "zygote" etymologically reminds one of the cell's origins in egg-joined-to-sperm; "totipotency" describes what it is capable of. A fertilized egg is precisely a "totipotent" cell; the product of human SCNT is, we assume, its equivalent.

In some discussions, the next few stages of the developing cloned human entity have been described as "clumps of cells." Yet, for reasons already given, this is only partially accurate. Viewed externally, under the microscope, the developing embryo will appear as two, then four, then eight cells "clumped" together, and the 100-to-200-cell blastocyst stage will indeed appear as a "ball of cells." Yet there is more here than meets the eye, for the "clump" is governed by an internal principle of development that shapes and directs its transformations. Thus, this ball or clump is not a mere heap or aggregate; it is a primordial and unfolding whole that functions as a whole and that is in the process of developing (or attempting to develop) into a mature whole being. Of course, if development is not pursued or not allowed to happen because of disruption, then the "clump of cells" description may be rendered accurate not just microscopically but also biologically. But as long as development continues and the developing entity is intact, that is not the case.

It would seem, then, that – whatever the reason for producing it – the initial product of somatic cell nuclear transfer is a living (one-celled) cloned human embryo. The *immediate* intention of transferring the nucleus is precisely to produce just such an entity: one that is alive (rather than nonliving), one that is human (rather than nonhuman or animal), and one that is an embryo, an entity capable of developing into an articulated organismic whole (rather than just a somatic cell capable only of replication into more of the same cell type). This is the intended primary product of performing SCNT, whether the ultimate motive or purpose is producing a live-born child from the cloned embryo or conducting scientific research on the cloned embryo. Also, the blastocyst stage that develops from this one-celled cloned embryo will be the same being, whether it is then transferred to a woman's uterus to begin a pregnancy or is used as a source of stem cells for research and possible therapy for others.

Yet, not surprisingly, objections have been raised to calling this cloned entity an "embryo," objections having to do both with its origins and with the uncertainty about the extent of its developmental potential. There are also objections having to do not with the facts but with public connotations and perceptions: for some members of the public, the word "embryos" apparently conjures images of miniature babies. If "nuclear transplantation to produce stem cells" seems to some people to be unfairly morally *neutered* terminology, "embryo" seems to other people to be unfairly morally *loaded* terminology, especially when used to describe an entity barely visible to the naked eye. We acknowledge this problem and recognize that, despite our best efforts, such difficulties in public perception probably cannot be simply corrected. But we do not regard this as sufficient reason to scrap the use of a term if it is in fact most appropriate. The other objections to calling the product of SCNT an "embryo" are not about rhetoric and politics, but about the thing itself. They should be addressed.

First, "human embryo," in the traditional scientific definition of this term, refers to the earliest stages of human development, from the zygote through roughly eight weeks of gestation, after which time it is called a fetus. Because the product of SCNT is technically not a zygote, not having come from egg and sperm, it is argued that it cannot therefore be an embryo. Second, it is said that it cannot be an embryo because it is an "artifact," something produced entirely by human artifice, "made" rather than "begotten." Third, we do not yet know for sure whether this entity can in fact develop into a baby; hence, we do not *know* whether it has the full developmental potential of a human embryo formed by fertilization.

There are, however, good responses to these objections. The first product of SCNT is, on good biological grounds, quite properly regarded as the *equivalent* of a zygote, and its subsequent stages as *embryonic stages in development*. True, it is not technically "zygotic" in origin, and it owes its existence to human artifice. But these objections, dealing only with *origins*, ignore the organization and powers of this entity, and the crucially important fact of its capacity to undergo future embryological development – just like a sexually produced embryo. True, it originates as a result of human artifice, and it lacks the natural bi-parental (male-plus-female) precursors. But this particular "artifact" is alive and self-developing, and should it eventually give rise to a baby, that child would in its being and its capacities be indistinguishable from any other human being – hardly an artifact – in the same way that Dolly is a sheep. True, regarding its developmental potential, we do not yet have incontrovertible proof that a cloned human embryo can in fact do what embryos are "supposed" to do and what animal cloned embryos have already done, namely, develop into all the later stages of the organism, up to its full maturity (à la Dolly). But if we do not assume this last possibility – an assumption based on the biological continuity of all mammals – there would be nothing to talk about in this whole matter of human cloning. As we emphasized in the first chapter of this report, *this entire inquiry assumes that cloned human embryos can someday be developed into live-born human beings.*

Once we make this assumption, neither its artificial nor its uni-parental source alters the decisive point: the product of SCNT is an entity that is the first stage of a developing organism – of a determinate species (human), with a full genetic complement, and its own (albeit near-replicated) individual genetic identity. It hence deserves on *functional grounds* to be called an *embryo*. And that is the heart of the reason why we in this report shall call it an "embryo" (actually, for reasons soon to be discussed, a "*cloned* embryo"): because the decisive questions to be addressed in our moral reflections have to do not with the origin of the entity but with its *developmental potential*, its embryonic character must be kept centrally in mind.

This decision, based on what we believe comes closest to the truth about the product of SCNT, is supported by other, more practical considerations. We are disinclined to introduce other words to describe the early product of human cloning that might deprive discussion of the ethics of human cloning of its humanly significant context. Despite the novelty of cloning and its products, their considerable kinship to elements of normal reproduction and development means that we enter upon the discussion equipped with existing and relevant terms and notions. We do not start in a terminological vacuum or with an empty dictionary. We observe that even people who prefer not to call the one-celled product of SCNT a zygote or embryo use terms like "blastocyst" and "embryo" to name the product a few cell divisions later. We think that using or coining other words will be more confusing to members of the public as they try to follow and contribute to the ethical discussion. And we clearly assume, as already stated, that the product of human SCNT could someday be shown to be capable of developing into a later-stage embryo, fetus, or live human being, even though such capacity has yet to be documented.

There are also very important *ethical* reasons that support our choice. We want to be very careful not to make matters easy for ourselves. We do not want to define away the moral questions of cloning-for-biomedical-research by denying to the morally crucial element a name that makes clear that there is a moral question to be faced. Yes, there is some ground for uncertainty about the being of the product of SCNT. Yet because something is ambiguous to *us* does not mean that it is ambiguous *in itself*. Where the moral stakes are high, we should not allow our uncertainty to lead us to regard the subject in question as being anything less than it might truly be.

The product of "SCNT" is not only an embryo; it is also a *clone*, genetically virtually identical to the individual that was the source of the transferred nucleus, hence an embryonic clone of the donor. There is, to be sure, much discussion about how close the genetic relation is between donor and embryonic clone, and about the phenotypic similarity of the clone to the donor. Yet the goal in this process is in fact a blastocyst-stage cloned embryo (in the case of cloning-for-biomedical-research) or a child who is genetically virtually identical to the donor (in the case of cloning-to-produce-children); otherwise there would be no reason to produce a cloned embryo by SCNT rather than an (uncloned) embryo by ordinary IVF. A full and fitting name of the developing entity produced by human SCNT is "cloned human embryo," a term that also allows us to remember that, thanks to its peculiar origins, this embryo is not in all respects identical to an embryo produced by fertilization of egg by sperm.

As if things were not difficult enough, a further complication may soon arise, following reports of successful SCNT experiments in which human somatic cells were fused with *animal* oocytes, and the resulting product grown to the blastocyst stage of development. What are we to call the product of this kind of cloning? And what kind of species identity does it have? According to the advance reports (based on a presentation at a scientific meeting), the stem cells extracted from the blastocyst stage were demonstrated to be human stem cells (somewhat surprisingly, the mitochondria were also human in genotype). Is this, therefore, a cloned human embryo? The only test that could settle the question – implantation into a woman's uterus for attempted gestation to see if a human child results – cannot ethically even be contemplated without already assuming a positive answer. In the face of uncertainty, therefore, and lest we err by overconfidence, there is *prima facie* reason to include even these cross-species entities in the category of "cloned human embryos." (When we come to the ethical issues of cloning-for-biomedical-research, we can consider whether this *terminological* judgment is matched by an *ethical* one.)

CONCLUSION

None of the terms available to us is entirely trouble-free. Yet the foregoing analysis leads us to the following conclusion regarding the terms best descriptive of the facts of the matter:

Human cloning (what it is): The asexual production of a new human organism that is, at all stages of development, genetically virtually identical to a currently existing or previously existing human being.

Human cloning (how it is done): It would be accomplished by introducing the nuclear material of a human somatic cell (donor) into an oocyte (egg) whose own nucleus has been removed or inactivated, yielding a product that has a human genetic constitution virtually identical to the donor of the somatic cell. This procedure is known as "somatic cell nuclear transfer" (SCNT).

Human cloning (why it is done): This same activity may be undertaken for purposes of producing children or for purposes of scientific and medical investigation and use, a distinction represented in the popular discussion by the terms "reproductive cloning" and "therapeutic cloning." We have chosen instead to use the following designations:

Cloning-to-produce-children: Production of a cloned human embryo, formed for the (proximate) purpose of initiating a pregnancy, with the (ultimate) goal of producing a child who will be genetically virtually identical to a currently existing or previously existing individual.

Cloning-for-biomedical-research: Production of a cloned human embryo, formed for the (proximate) purpose of using it in research or for extracting its stem cells, with the (ultimate) goals of gaining scientific knowledge of normal and abnormal development and of developing cures for human diseases.

Cloned human embryo: (a) The immediate and developing product of the initial act of cloning, accomplished by SCNT. (b) A human embryo resulting from the somatic cell nuclear transfer process (as contrasted with a human embryo arising from the union of egg and sperm).

10

Scientific Background for Population Ecology

INTRODUCTION

The purpose of this chapter is to provide background on basic scientific aspects of human cloning for readers of this report. Background on stem cell research is also included to enable readers to understand how cloned embryos might be useful in stem cell and other biomedical research. This limited treatment only summarizes and highlights basic aspects of these topics, in part because two major detailed reports, *Scientific and Medical Aspects of Human Reproductive Cloning* and *Stem Cells and the Future of Regenerative Medicine*, have been recently published.

This review is based largely on scientific research papers published through June 2002, supplemented by references to several articles in the popular press. However, the research areas of cloning and stem cell research are being very actively investigated, and significant new developments are published frequently. Publication of new results could change some of the interpretations and emphases in this review.

Use of unfamiliar technical terms has been avoided wherever possible. Scientific names and terms used are described and defined in the Glossary of Terms.

Some Basic Facts about Human Cell Biology and Sexual Reproduction

We begin with some basic facts about human cells, germ cells (egg and sperm), and early embryonic development to provide the background for understanding the mechanism of cloning and the differences between sexual and asexual reproduction.

Normal human cells with nuclei contain forty-six chromosomes, twenty-two pairs plus two X chromosomes if the individual is female, or twenty-two pairs plus one X and one Y chromosome if the individual is male. These chromosomes contain nearly all of the cell's DNA and, therefore, the genes of the cell. During formation of sperm cells, a process of specialized cell division produces mature sperm cells containing twenty-three chromosomes (twenty-two unpaired chromosomes plus either X or Y). During the formation of eggs (oocytes), a process of specialized cell division produces a cell containing two pronuclei, each of which contains twenty-two unpaired chromosomes plus an X. During fertilization, a polar body containing one of these pronuclei is ejected from the egg.

Fusion of egg and sperm cells and the subsequent fusion of their nuclei (the defining acts of all sexual reproduction) produce a zygote that again contains a nucleus with the adult cell complement of forty-six chromosomes, half from each parent [See Figure 1]. The zygote then begins the gradual process of cell division, growth, and differentiation. After four to five days, the developing embryo attains the 100-200 cell (blastocyst) stage. In normal reproduction, the blastocyst implants into the wall of the uterus, where, suitably nourished, it continues the process of coordinated cell, tissue, and organ differentiation that eventually produces the organized, articulated, and integrated whole that is the newborn infant.

According to some estimates, about half of all early human embryos fail to implant, and are expelled with the menses during the next menstrual cycle.

Not quite all the DNA of a human cell resides in its nucleus. All human cells, including eggs and sperm, contain small, energy-producing organelles called mitochondria. Mitochondria contain a small piece of DNA that specifies the genetic instructions for making several essential mitochondrial proteins. When additional mitochondria are produced in the cell, the mitochondrial DNA is replicated, and a copy of it is passed along to the new mitochondria that are formed. During fertilization, sperm mitochondria are selectively degraded inside the zygote. Thus, the developing embryo inherits solely or principally mitochondria (and mitochondrial DNA) from the egg.

Human reproduction has also been accomplished with the help of in vitro fertilization (IVF) of eggs by sperm, and the subsequent transfer of one or more early embryos to a woman for gestation and birth. Even though such union of egg and sperm requires laboratory assistance and takes place outside of the body, human reproduction using IVF is still sexual in the biological sense: the new human being arises from two biological parents through the union of egg and sperm.

Egg and sperm cells combined in vitro have also been used to start the process of animal development. Transfer of the resulting blastocysts into the uterus of a female of the appropriate animal species is widely used in animal husbandry with resulting successful live births.

CLONING (ASEXUAL REPRODUCTION) OF MAMMALS

The startling announcement that Dolly the sheep had been produced by cloning indicated that it was possible to produce live mammalian offspring via asexual reproduction through cloning with adult donor cell nuclei. In outline form, the steps used to produce live offspring in the mammalian species that have been cloned so far are:

1. Obtain an egg cell from a female of a mammalian species.
2. Remove the nuclear DNA from the egg cell, to produce an enucleated egg.
3. Insert the nucleus of a donor adult cell into the enucleated egg, to produce a reconstructed egg.
4. Activate the reconstructed egg with chemicals or electric current, to stimulate the reconstructed egg to commence cell division.
5. Sustain development of the cloned embryo to a suitable stage in vitro, and then transfer the resulting cloned embryo to the uterus of a female host that has been suitably prepared to receive it.

6. Bring to live birth a cloned animal that is genetically virtually identical (except for the mitochondrial DNA) to the animal that donated the adult cell nucleus.

Cloning to produce live offspring carries with it several possibilities not available through sexual reproduction. Because the number of presumably identical donor cells is very large, this process could produce a very large number of genetically virtually identical individuals, limited only by the supply of eggs and female animals that could bear the young. In principle, any animal, male or female, newborn or adult, could be cloned, and in any quantity.

Because mammalian cells can be frozen and stored for prolonged periods at low temperature and grown again for use as donor cells in cloning, one may even clone individuals who have died. In theory, a clone could be cloned again, on and on, without limit. In mice, such "cloning of clones" has extended out to six generations.

Why is production of live cloned mammalian offspring a relatively rare event? Several factors may play a role. Enucleation of the egg may (variably from one attempt to the next) remove or damage its "epigenetic reprogramming" capabilities. Isolating a nucleus from the donor cell and manipulating it to insert it into the egg is also a traumatic process that may damage the nucleus. An optimal in vitro nutritive environment for the development of cloned animal embryos may not yet have been determined. One interpretation attributes the early death of many cloned embryos to complete failure or incompleteness of epigenetic reprogramming.

EPIGENETIC MODIFICATION AND REPROGRAMMING

Normal mammalian embryonic development results from selective expression of some genes and repression of others. Tissue differentiation depends upon several types of "epigenetic modifications" of DNA structure and spatial organization that selectively turn genes on and off. The chromosomal DNAs of egg and sperm cells are modified during their maturation, so that at fertilization, both sets of DNA are ready for the complex pattern of gene expression required for normal embryonic development.

In order for the DNA of a differentiated *adult* cell to direct embryonic development in cloning, it must be "epigenetically reprogrammed." That is, the epigenetic modifications that allowed the cell to express genes appropriate for, for example, a differentiated skin cell must be reduced, and the gene expression program required for full embryonic development must be activated.

The completeness of epigenetic reprogramming is crucial for successful cloning-to-produce-children. It will also be important to assess the impact of variation in epigenetic reprogramming on the biological properties of cloned stem cell preparations. If the extent of epigenetic reprogramming varies from one cloning event to the next, the protein expression pattern and thus the biological properties of cloned stem cell preparations may also vary. Thus, it may be necessary to produce and test multiple cloned stem cell preparations before preparations that are informative about human disease or useful in cellular transplantation therapies can be identified.

CLONING-TO-PRODUCE-CHILDREN

At this writing, it is uncertain whether anyone has attempted cloning-to-produce-children. Although claims of such attempts have been reported in the press, no credible evidence of any such experiments has been reported as of June 2002. Thus, it is not yet known whether a transferred cloned human embryo can progress all the way to live birth. However, the steps in such an experiment would probably be similar to those described for animal cloning. After a thorough review of the data on animal cloning, the NAS panel, in its report *Scientific and Medical Aspects of Human Cloning* [page ES-1], came to the following conclusion: "It [cloning-to-produce-children] is dangerous and likely to fail."

STEM CELLS AND REGENERATIVE MEDICINE

The subject of stem cell research is much too large to be covered extensively here. Yet the following information on stem cells and their possible uses in medical treatments should facilitate understanding of the relationships between cloning-for-biomedical-research and stem cells.

Stem cells are undifferentiated multipotent precursor cells that are capable both of perpetuating themselves as stem cells and of undergoing differentiation into one or more specialized types of cells (for example, kidney, muscle). Human embryonic stem cells have been isolated from embryos at the blastocyst stage or from the germinal tissue of fetuses. Multipotent adult progenitor cells have been isolated from sources such as human and rodent bone marrow. Such cell populations can be differentiated in vitro into a number of different cell types, and thus are the subject of much current research into their possible uses in regenerative medicine. Cloned human embryonic stem cell preparations could be produced using somatic cell nuclear transfer to produce a cloned human embryo, and then taking it apart at the (100-200 cell) blastocyst stage and isolating stem cells (see Figure 2). These stem cells would be genetically virtually identical to cells from the nucleus donor.

Scientists are pursuing the development of therapies based on transplantation of cells for several human diseases, including Parkinson's disease and Type I diabetes. In Parkinson's disease, particular brain cells that produce the essential neurotransmitter dopamine die selectively. Experimental clinical treatment involving transplantation of human *fetal* brain cell populations, in which a small fraction of the cells produce dopamine, has improved the condition of some Parkinson's disease patients. Dopamine-producing neurons derived from mouse embryonic stem cells have been shown to function in an animal model of Parkinson's disease. Thus, there is a possibility that transplantation of dopamine-producing neural cells derived from embryonic or adult stem cell populations might be a useful treatment for Parkinson's disease in the future.

However, to be effective as long-term treatments of Parkinson's disease, Type I diabetes, and other diseases, cell transplantation therapies will have to overcome the immune rejection problem. Cells from one person transplanted into the body of another are usually recognized as foreign and killed by the immune system. If cells derived from stem cell preparations are to be broadly useful in transplantation therapies for human diseases, some way or ways around this problem will have to be found. For example, if the cells were isolated from a cloned human embryo at the blastocyst stage, in which the donor nucleus came from a patient with Parkinson's disease, in theory these stem cells would produce the same proteins as the patient. The hope is that dopamine-producing cells derived from these "individualized" stem cell preparations would not be immunologically rejected upon transplantation back into the Parkinson's disease patient. Alternatively, if dopamine-producing cells could be derived from the patient's own adult stem cell or multipotent adult precursor cell populations, they could also be used in such therapies. Another possibility is mentioned in a press report about work with a single Parkinson's disease patient, in which brain cells were removed from the patient, expanded by growth in vitro, stimulated to increase dopamine production, and transplanted back into the brain of the same patient with an observed reduction in disease symptoms.

By combining specific gene modification and cloned stem cell procedures, Rideout et al. have provided a remarkable example of how some human genetic diseases might someday be treated. Starting with a mouse strain that was deficient in immune system function because of a gene mutation, these investigators (1) produced a cloned stem cell line carrying the gene mutation, (2) specifically repaired the gene mutation in vitro, (3) differentiated the repaired cloned stem cell preparation in vitro into bone marrow precursor cells, and (4) treated the mutant mice with the repaired bone marrow precursor cells and observed a restoration of immune cell function.

Although remarkably successful, the experimental results included a caveat. The investigators also observed a tendency of even these cloned bone marrow precursor cells to be recognized as foreign by the recipient mice. Rideout et al. were led to conclude: "Our results raise the provocative possibility that even genetically matched cells derived by therapeutic cloning may still face barriers to effective transplantation for some disorders."

Lanza et al. have also evaluated the potential for immune rejection of cloned embryonic materials, while showing the potential therapeutic value of tissues taken from cloned fetuses. Cloned cattle embryos at the blastocyst stage were transferred to the uteri of surrogate mothers and allowed to develop for five to eight weeks. Fetal heart, kidney, and skeletal muscle tissues were isolated, and degradable polymer vehicles containing these cloned cells were then transplanted back into the animals that donated the nuclei for cloning.

The investigators observed no rejection reaction to the transplanted cloned cells using two different immunological tests. More investigations with cloned stem cell materials involving different stem cell preparations of varying sizes, different sites of implantation, and sensitive tests to detect low levels of immunological rejection will be required for a complete assessment of the possibility of using cloned stem cell populations to solve the immune rejection problem.

HUMAN CLONING-FOR-BIOMEDICAL-RESEARCH

Producing cloned stem cell preparations for possible use in individual patients suffering from diseases like Parkinson's disease and Type I diabetes is one reason to pursue cloning-for-biomedical-research. In vitro production of cloned human embryos could also be important to scientists interested in studying early human development. Stem cells derived from cloned human embryos at the blastocyst stage that were produced with nuclei from individuals with genetic diseases could be useful in the study of the critical events that lead to these diseases (for example, see Bahn et al.). Specific genes could be introduced into developing human embryos to obtain information about the role or roles of these genes in early human development.

One attempt at human cloning-for-biomedical-research has been published in the scientific literature by Cibelli et al. as of the end of June 2002. It involved the following steps:

1. Obtain human eggs from informed and consenting female volunteers.
2. Remove the nuclear DNA from the egg cell, to produce an enucleated egg.
3. Insert the nucleus of a cell from an informed and consenting adult donor into the enucleated egg, to produce a reconstituted egg.

4. Activate the reconstituted egg with chemicals or electricity to stimulate it to commence cell division in vitro, producing a cloned embryo.
5. Use a microscope to follow the early cell divisions of the cloned embryo.

In the experiments described by Cibelli et al., the stated intent was to create cloned human embryos that would progress to the 100-200 cell stage, at which point the cloned embryo would be taken apart, stem cells would be isolated from the inner cell mass, and an attempt would be made to grow and preserve "individualized" human stem cells for the possible future medical benefit of the somatic cell donor. Because the cloned human embryos stopped dividing and died at the six-cell stage, no stem cells were isolated in these experiments. In light of results in other animal species and the variable completeness of "epigenetic reprogramming," it is perhaps not surprising that sixteen of the nineteen cloned human embryos described by Cibelli et. al. did not undergo cell division and none of the other three divided beyond the six-cell stage.

Although the steps these researchers followed in these experiments were the same as those that would be used by those attempting human cloning-to-produce-children, they distinguished their intent from such cloning by stating: "Strict guidelines for the conduct of this research have been established by Advanced Cell Technology's independent Ethics Advisory Board (EAB). In order to prevent any possibility of reproductive cloning, the EAB requires careful accounting of all eggs and embryos used in the research. No embryo created by means of NT [nuclear transfer] technology may be maintained beyond 14 days of development."

PARTHENOGENESIS (ANOTHER FORM OF ASEXUAL REPRODUCTION)

Using chemical or electrical stimuli, it is also possible to stimulate human eggs to undergo several rounds of cell division, as if they had been fertilized. In this case, the egg retains all forty-six egg cell chromosomes and egg cell mitochondria. In amphibians, this asexual reproduction process, known as parthenogenesis, has produced live offspring that contain the same nuclear DNA as the egg. These offspring are all necessarily female. Parthenogenesis in mammals has not led reproducibly to the production of live offspring.

Cibelli et al. activated human eggs (obtained from informed and consenting donors) by parthenogenesis, and obtained multiple cell divisions up to the early embryo stage in six out of twenty-two attempts. Although there was no report that stem cells were isolated in these experiments, it is possible that parthenogenesis of human eggs could induce them to develop to a stage where parthenogenetic stem cells could be isolated. For example, Cibelli et al. derived a monkey parthenogenetic stem cell preparation from Macaca fasicularis eggs activated by parthenogenesis. Whether cloned stem cells resulting from parthenogenesis have been completely and correctly epigenetically reprogrammed remains to be determined.

11

The Ethics of Cloning-to-Produce-Children

Cloning-to-produce-children has been the subject of two major national reports in recent years – first by the National Bioethics Advisory Commission in June 1997, and more recently by the National Academy of Sciences in January 2002. Both reports concluded that attempts to clone a human being "at this time" would be unethical, owing to questions about the safety of the technique and the likelihood of physical harm to those involved. But both reports also concluded that the nation required much deeper reflection about the "ethical and social implications" of cloning-to-produce-children beyond the scientific and medical aspects of the procedure. As the National Academy of Sciences report stated:

> Our present opposition to human reproductive cloning is based on science and medicine, irrespective of broader considerations. The panel stresses, however, that a broad ethical debate must be encouraged so that the public can be prepared to make decisions if human reproductive cloning is some day considered medically safe for mothers and offspring.

In this chapter we attempt to take up this charge to engage in a broad ethical consideration of the merits of cloning-to-produce-children.

The prospect of cloning-to-produce-children raises a host of moral questions, among them the following: Could the first attempts to clone a human child be made without violating accepted moral norms governing experimentation on human subjects? What harms might be inflicted on the cloned child as a consequence of having been made a clone? Is it significant that the cloned child would inherit a genetic identity lived in advance by another – and, in some cases, the genetic identity of the cloned child's rearing parent? Is it significant that cloned children would be the first human beings whose genetic identity was entirely known and selected in advance?

How might cloning-to-produce-children affect relationships within the cloning families? More generally, how might it affect the relationship between the generations? How might it affect the way society comes to view children? What other prospects would we be tacitly approving in advance by accepting this practice? What important human goods might be enhanced or sacrificed were we to approve cloning-to-produce-children?

In what follows, we shall explicitly consider many of these questions. But as we do so, we shall not lose sight of the larger and fundamental human contexts discussed in Chapter One – namely, the meaning of human procreation and care of children, the means and ends of biotechnology, and the relation between science and society.

Indeed, overarching our entire discussion of the *specific* ethical issues is our concern for the human significance of procreation as a whole and our desire to protect what is valuable in it from erosion and degradation – not just from cloning but from other possible technological and nontechnological dangers. Readers of this report are encouraged to consider the discussion that follows in a similar light.

We will begin by formulating the best moral case for cloning-to-produce-children – describing both the specific purposes it might serve and the philosophic and moral arguments made in its favor. From there we will move to the moral case against cloning-to-produce-children. Beginning with the safety objections that have dominated the debate thus far, we will show how these concerns ultimately point beyond themselves toward broader ethical concerns. Chief among these is how cloning-to-produce-children would challenge the basic nature of human procreation and the meaning of having children. We shall also consider cloning's effects on human identity, how it might move procreation toward a form of manufacture or toward eugenics, and how it could distort family relations and affect society as a whole.

THE CASE FOR CLONING-TO-PRODUCE-CHILDREN

Arguments in defense of cloning-to-produce-children often address questions of reproduction, but they tend to focus on only a relatively narrow sliver of the goods and principles involved. This certainly does not mean that such arguments lack merit. Indeed, some of the arguments in favor of cloning-to-produce-children appeal to the deepest and most meaningful of our society's shared values.

Purposes

In recent years, in anticipation of cloning-to-produce-children, proponents have articulated a variety of possible uses of a perfected technology: providing a "biologically related child" for an infertile couple; permitting reproduction for single individuals or same-sex couples; avoiding the risk of genetic disease; securing a genetically identical source of organs or tissues perfectly suitable for transplantation; "replacing" a loved spouse or child who is dying or has died; obtaining a child with a genotype of one's own choosing (including one's own genotype); replicating individuals of great genius, talent, or beauty, or individuals possessing traits that are for other reasons attractive to the cloners; and creating sets of genetically identical humans who might have special advantages in highly cooperative ventures in both war and peace.

The desire to control or select the genomes of children-to-be through cloning has charmed more than a few prospective users, in the United States and around the world.

Although we appreciate that a perfected technology, once introduced for one purpose, might then be used for any of these purposes, we shall examine further only those stated purposes that seem to us to merit serious consideration.

To Produce Biologically Related Children

Human cloning would allow individuals or couples with fertility problems to have biologically related children. For example, if a man could not produce sperm, cloning would allow him to have a child who is "biologically related" to him. In addition, it would allow married couples with fertility problems to avoid using donor gametes, and therefore avoid raising children with genetic inheritances from outside the marriage.

To Avoid Genetic Disease

Human cloning could allow couples at risk of generating children with genetic disease to have healthy children. For example, if both parents carried one copy of a recessive gene for the same heritable disorder, cloning might allow them to ensure that their child does not inherit the known genetic disease (without having to resort to using donor gametes or practicing preimplantation or prenatal genetic diagnosis and elimination of afflicted embryos or fetuses).

To Obtain "Rejection-Proof" Transplants

Human cloning could produce ideal transplant donors for people who are sick or dying. For example, if no genetic match could be found for a sick child needing a kidney or bone marrow transplant, and the parents had planned to have another child, cloning could potentially serve the human goods of beginning a new life and saving an existing one.

To "Replicate" a Loved One

Human cloning would allow parents to "replicate" a dead or dying child or relative. For example, one can imagine a case in which a family – mother, father, and child – is involved in a terrible car accident in which the father dies instantly and the child is critically injured. The mother, told that her child will soon die, decides that the best way to redeem the tragedy is to clone her dying child. This would allow her to preserve a connection with both her dead husband and her dying child, to create new life as a partial human answer to the grievous misfortune of her child's untimely death, and to continue the name and biological lineage of her deceased husband.

To Reproduce Individuals of Great Genius, Talent, or Beauty

Human cloning would allow families or society to reproduce individuals of great genius, talent, or beauty, where these traits are presumed to be based on the individuals' desirable or superior genetic makeups. For example, some admirers of great athletes, musicians, or mathematicians, believing that the admired attributes are the result of a superior genetic endowment, might want to clone these distinguished individuals. Just as the cloning of cattle is being promoted as a means of perpetuating champion milk- or meat-producing cows, so cloning-to-produce-children has been touted as a means of perpetuating certain "superior" human exemplars.

Arguments

The purposes or reasons for cloning-to-produce-children are, as they are stated, clearly intelligible on their face. When challenged, the defenders of these purposes often appeal to larger moral and political goods. These typically fall within the following three categories: human freedom, existence, and well-being.

The Goodness of Human Freedom

Strictly speaking, the appeal to human freedom is not so much a defense of cloning itself as it is of the *right* to practice it, asserted against those who seek to prohibit it. No one, we suspect, would say that he wanted to clone himself or any one else in order to be free or to vindicate the goodness of liberty. Nevertheless, human freedom is a defense often heard in support of a "right" to clone.

Those who defend cloning-to-produce-children on the grounds of human freedom make two kinds of arguments. The first is that because individuals in pluralistic societies have different definitions of the good life and of right and wrong, society must protect individual freedom to choose against the possible tyranny of the majority. This means securing and even expanding the rights of individuals to make choices so long as their choices do not directly infringe on the rights (and especially the physical safety) of other rights-bearing citizens. In *Eisenstadt v. Baird* (1972), the United States Supreme Court enunciated what has been called a principle of reproductive freedom: "If the right to privacy means anything, it is the right of the individual, married or single, to be free from unwarranted governmental intrusion into matters so affecting a person as a decision whether to bear or beget a child." Defenders of cloning-to-produce-children argue that, in the event that the physical risks to mother and future child were shown to be ethically acceptable, the use of this new reproductive technology would fall under the protective umbrella of reproductive freedom.

A second defense of human cloning on the grounds of freedom is the claim that human existence is by its very nature "open-ended," "indeterminate," and "unpredictable." Human beings are always remaking themselves, their values, and their ways of interacting with one another. New technologies are central to this open-ended idea of human life, and to shut down such technologies simply because they change the "traditional" ways of doing things is unjustifiable.

As constitutional scholar Laurence Tribe has argued in reference to human cloning: "A society that bans acts of human creation that reflect unconventional sex roles or parenting models (surrogate motherhood, in vitro fertilization, artificial insemination, and the like) for no better reason than that such acts dare to defy 'nature' and tradition (and to risk adding to life's complexity) is a society that risks cutting itself off from vital experimentation and risks sterilizing a significant part of its capacity to grow."

The Goodness of Existence

Like the appeal to freedom, the appeal to the goodness of existence is not an argument *for* cloning, but an argument *against* opponents who speak up in the name of protecting the cloned child-to-be against the harms connected with its risky and strange origins as a clone. This argument asserts that attempts to produce children through cloning, like *any* attempt to produce a child, will directly benefit the cloned child-to-be, since without the act of cloning the child in question would not exist. Existence itself, it is argued, is the first "interest" that makes all other interests – including the interests of safety and well-being – possible. Even taking into account the possibility of serious genetic or developmental disorders, this position holds that a cloned individual, once born, would prefer existence as a clone to no existence at all. There is also a serious corollary about how, in the absence of a principle that values existence *as such*, we will and should regard and treat people born with disabilities or deformities:

opponents of cloning might appear in a position of intolerance – of saying to cloned individuals, "Better for us (and for you) had you never existed."

The Goodness of Well-Being

The third moral argument for cloning-to-produce-children is that it would contribute in certain cases to the fulfillment of human goods that are widely honored and deeply rooted in modern democratic society. These human goods include the health of newborn and existing children, reproductive possibilities for infertile couples, and the possibility of having a biologically related child. In all these circumstances, human cloning could relieve existing suffering and sorrow or prevent them in the future. Those who take this position do not necessarily defend human cloning-to-produce-children as such. Rather, they argue that a moral and practical line can be drawn between cloning-to-produce-children that serves the "therapeutic" aims of health (for the cloned child-to-be, for the infertile couple, or for an existing child) and the "eugenic" aims of producing or mass-producing superior people.

Some people argue more broadly that an existing generation has a responsibility to ensure, to the extent possible, the genetic quality and fitness of the next generation. Human cloning, they argue, offers a new method for human control and self-improvement, by allowing families to have children free of specific genetic diseases or society to reproduce children with superior genetic endowments. It also provides a new means for gaining knowledge about the age-old question of nature versus nurture in contributing to human achievement and human flourishing, and to see how clones of great geniuses measure up against the "originals."

Critique and Conclusion

While we as a Council acknowledge merit in some of the arguments made for cloning-to-produce-children, we are generally not persuaded by them. The fundamental weakness of the proponents' case is found in their incomplete view of human procreation and families, and especially the place and well-being of children. Proponents of cloning tend to see procreation primarily as the free exercise of a parental right, namely, a right to satisfy parental desires for self-fulfillment or a right to have a child who is healthy or "superior." Parents seek to overcome obstacles to reproduction, to keep their children free of genetic disease or disorder, and to provide them with the best possible genetic endowment. The principles guiding such prospective parents are freedom (for themselves), control (over their child), and well-being (both for themselves and what they imagine is best for their child). Even taken together, these principles provide at best only a partial understanding of the meaning and entailments of human procreation and child-rearing. In practice, they may prove to undermine the very goods that the proponents of cloning aim to serve, by undermining the unconditional acceptance of one's offspring that is so central to parenthood.

There are a number of objections – or at the very least limitations – to viewing cloning-to-produce-children through the prism of rights. Basic human rights are usually asserted on behalf of the human individual agent: for example, a meaningful right not to be prevented from bearing a child can be asserted for each individual against state-mandated sterilization programs. But the act of procreation is not an act involving a single individual. Indeed, until human cloning arrives, it continues to be impossible for any one person to procreate alone. More important, there is a crucial third party involved: the child, whose centrality to the activity exposes the insufficiency of thinking about procreation in terms of rights.

After all, rights are limited in the following crucial way: they cannot be ethically exercised at the expense of the rights of another. But the "right to reproduce" cannot be ethically exercised without at least considering the child that such exercise will bring into being and who is at risk of harm and injustice from the exercise. This obligation cannot be waived by an appeal to the absolutist argument of the goodness of existence. Yes, existence is a primary good, but that does not diminish the ethical significance of knowingly and willfully putting a child in grave physical danger in the very act of giving that child existence. It is certainly true that a life with even severe disability may well be judged worth living by its bearer: "It is better to have been born as I am than not to be here at all." But if his or her disability was caused by behavior that could have been avoided by parents (for example, by not drinking or using drugs during pregnancy, or, arguably, by not cloning), many would argue that they should have avoided it. A post-facto affirmation of existence by the harmed child would not retroactively excuse the parental misconduct that caused the child's disability, nor would it justify their failure to think of the child's well-being as they went about exercising their "right to procreate." Indeed, procreation is, by its very nature, a limitation of absolute rights, since it brings into existence another human being toward whom we have responsibilities and duties.

In short, the right to decide "*whether* to bear or beget a child" does not include a right to have a child *by whatever means*. Nor can this right be said to imply a corollary – the right to decide what kind of child one is going to have. There are at least some circumstances where reproductive freedom must be limited to protect the good of the child (as, for instance, with the ban on incest). Our society's commitment to freedom and parental authority by no means implies that all innovative procedures and practices should be allowed or accepted, no matter how bizarre or dangerous.

Proponents of cloning, when they do take into account the interests of the child, sometimes argue that this interest justifies and even requires thoroughgoing parental control over the procreative process. Yet this approach, even when well-intentioned, may undermine the good of the child more than it serves the child's best interests. For one thing, cloning-to-produce-children of a desired or worthy sort overlooks the need to restrain the parental temptation to total mastery over children. It is especially morally dubious for this project to go forward when we know so little about the unforeseen and unintended consequences of exercising such genetic control. In trying by cloning to circumvent the risk of genetic disease or to promote particular traits, it is possible – perhaps likely – that new risks to the cloned child's health and fitness would be inadvertently introduced (including the forgoing of genetic novelty, a known asset in the constant struggle against microbial and parasitic diseases). Parental control is a double-edged sword, and proponents seem not to acknowledge the harms, both physical and psychological, that may befall the child whose genetic identity is selected in advance. The case for cloning in the name of the child's health and well-being is certainly the strongest and most compelling.

The desire that one's child be free from a given genetic disease is a worthy aspiration. We recognize there may be some unusual or extreme cases in which cloning might be the best means to serve this moral good, if other ethical obstacles could somehow be overcome. (A few of us also believe that the desire to give a child "improved" or "superior" genetic equipment is not necessarily to be condemned.)

However, such aspirations could endanger the personal, familial, and societal goods supported by the character of human procreation. We are willing to grant that there may be exceptional cases in which cloning-to-produce-children is morally defensible; however, that being said, we would also argue that such cases do not justify the harmful experiments and social problems that might be entailed by engaging in human cloning. Hard cases are said to make bad law. The same would be true for succumbing to the rare, sentimentally appealing case in which cloning seems morally plausible.

Finally, proponents do not adequately face up to the difficulty of how "well-being" is to be defined. Generally, they argue that these matters are to be left up to the free choices of parents and doctors. But this means that the judgments of "proper" and "improper" will be made according to subjective criteria alone, and under such circumstances, it will be almost impossible to rule out certain "improvements" as unacceptable.

In the sections that follow, we shall explain more fully why Members of the Council are not convinced by the arguments for cloning-to-produce-children, even in the most defensible cases. To see why this is so, we need to consider cloning-to-produce-children from the broadest possible moral perspective, beginning with ethical questions regarding experiments on human subjects. What we hope to show is that the frequently made safety arguments strike deeper than we usually realize, and that they point beyond themselves toward more fundamental moral objections to cloning-to-produce-children.

THE CASE AGAINST CLONING-TO-PRODUCE-CHILDREN

The Ethics of Human Experimentation

We begin with concerns regarding the safety of the cloning procedure and the health of the participants. We do so for several reasons. First, these concerns are widely, indeed nearly unanimously, shared. Second, they lend themselves readily to familiar modes of ethical analysis – including concerns about harming the innocent, protecting human rights, and ensuring the consent of all research subjects. Finally, if carefully considered, these concerns begin to reveal the important ethical principles that must guide our broader assessment of cloning-to-produce-children. They suggest that human beings, unlike inanimate matter or even animals, are in some way *inviolable*, and therefore challenge us to reflect on what it is *about* human beings that makes them inviolable, and whether cloning-to-produce-children threatens these distinctly human goods.

In initiating this analysis, there is perhaps no better place to start than the long-standing international practice of regulating experiments on human subjects. After all, the cloning of a human being, as well as all the research and trials required before such a procedure could be expected to succeed, would constitute experiments on the individuals involved – the egg donor, the birthing mother, and especially the child-to-be. It therefore makes sense to consider the safety and health concerns that arise from cloning-to-produce-children in light of the widely shared ethical principles that govern experimentation on human subjects.

Since the Second World War, various codes for the ethical conduct of human experimentation have been adopted around the world. These codes and regulations were formulated in direct response to serious ethical lapses and violations committed by research scientists against the rights and dignity of individual human beings. Among the most important and widely accepted documents to emerge were the Nuremberg Code of 1947 and the Helsinki Declaration of 1964. Influential in the United States is also the Belmont Report, published in 1978 by the National Commission for the Protection of Human Subjects of Biomedical and Behavioral Research.

The Nuremberg Code laid out ten principles for the ethical conduct of experiments, focusing especially on voluntary consent of research subjects, the principle that experiments should be conducted only with the aim of providing a concrete good for society that is unprocurable by other methods, and with the avoidance of physical or mental harm. The Helsinki Declaration stated, among other things, that research should be undertaken only when the prospective benefit clearly outweighs the expected risk, when the research subject has been fully informed of all risks, and when the research-subject population is itself likely to benefit from the results of the experiment.

Finally, the Belmont Report proposed three basic ethical principles that were to guide the treatment of human subjects involved in scientific research. The first of these is *respect for persons,* which requires researchers to acknowledge the autonomy and individual rights of research subjects and to offer special protection to those with diminished autonomy and capacity. The second principle is *beneficence.*

Scientific research must not only refrain from harming those involved but must also be aimed at helping them, or others, in concrete and important ways. The third principle is *justice,* which involves just distribution of potential benefits and harms and fair selection of research subjects. When applied, these general principles lead to both a requirement for informed consent of human research subjects and a requirement for a careful assessment of risks and benefits before proceeding with research. Safety, consent, and the rights of research subjects are thus given the highest priority.

It would be a mistake to view these codes in narrow or procedural terms, when in fact they embody society's profound sense that human beings are not to be treated as experimental guinea pigs for scientific research. Each of the codes was created to address a specific disaster involving research science – whether the experiments conducted by Nazi doctors on concentration camp prisoners, or the Willowbrook scandal in which mentally retarded children were infected with hepatitis, or the Tuskegee scandal in which underprivileged African-American men suffering from syphilis were observed but not treated by medical researchers –

and each of the codes was an attempt to defend the inviolability and dignity of all human beings in the face of such threats and abuses. More simply stated, the codes attempt to defend the weak against the strong and to uphold the equal dignity of all human beings. In taking up the application of these codes to the case of cloning-to-produce-children, we would suggest that the proper approach is not simply to discover specific places where human cloning violates this or that stipulation of this or that code, but to grapple with how such cloning offends the spirit of these codes and what they seek to defend.

The ethics of research on human subjects suggest three sorts of problems that would arise in cloning-to-produce-children: (1) problems of safety; (2) a special problem of consent; and (3) problems of exploitation of women and the just distribution of risk. We shall consider each in turn.

Problems of Safety

First, cloning-to-produce-children is not now safe. Concerns about the safety of the individuals involved in a cloning procedure are shared by nearly everyone on all sides of the cloning debate. Even most proponents of cloning-to-produce-children generally qualify their support with a caveat about the safety of the procedure. Cloning experiments in other mammals strongly suggest that cloning-to-produce-children is, at least for now, far too risky to attempt. Safety concerns revolve around potential dangers to the cloned child, as well as to the egg donor and the woman who would carry the cloned child to birth.

(a) Risks to the child. Risks to the cloned child-to-be must be taken especially seriously, both because they are most numerous and most serious and because – unlike the risks to the egg donor and birth mother – they cannot be accepted knowingly and freely by the person who will bear them. In animal experiments to date, only a small percentage of implanted clones have resulted in live births, and a substantial portion of those live-born clones have suffered complications that proved fatal fairly quickly. Some serious though nonfatal abnormalities in cloned animals have also been observed, including substantially increased birth-size, liver and brain defects, and lung, kidney, and cardiovascular problems.

Longer-term consequences are of course not known, as the oldest successfully cloned mammal is only six years of age. Medium-term consequences, including premature aging, immune system failure, and sudden unexplained death, have already become apparent in some cloned mammals. Some researchers have also expressed concerns that a donor nucleus from an individual who has lived for some years may have accumulated genetic mutations that – if the nucleus were used in the cloning of a new human life – may predispose the new individual to certain sorts of cancer and other diseases.

(b) Risks to the egg donor and the birth mother. Accompanying the threats to the cloned child's health and well-being are risks to the health of the egg donors. These include risks to her future reproductive health caused by the hormonal treatments required for egg retrieval and general health risks resulting from the necessary superovulation.

Animal studies also suggest the likelihood of health risks to the woman who carries the cloned fetus to term. The animal data suggest that late-term fetal losses and spontaneous abortions occur substantially more often with cloned fetuses than in natural pregnancies. In humans, such late-term fetal losses may lead to substantially increased maternal morbidity and mortality. In addition, animal studies have shown that many pregnancies involving cloned fetuses result in serious complications, including toxemia and excessive fluid accumulation in the uterus, both of which pose risks to the pregnant animal's health. In one prominent cattle cloning study, just under one-third of the pregnant cows died from complications late in pregnancy.

Reflecting on the dangers to birth mothers in animal cloning studies, the National Academy report concluded: Results of animal studies suggest that reproductive cloning of humans would similarly pose a high risk to the health of both fetus or infant and mother and lead to associated psychological risks for the mother as a consequence of late spontaneous abortions or the birth of a stillborn child or a child with severe health problems.

(c) An abiding moral concern. Because of these risks, there is widespread agreement that, at least for now, attempts at cloning-to-produce-children would constitute unethical experimentation

on human subjects and are therefore impermissible. These safety considerations were alone enough to lead the National Bioethics Advisory Commission in June 1997 to call for a temporary prohibition of human cloning-to-produce-children. Similar concerns, based on almost five more years of animal experimentation, convinced the panel of the National Academy of Sciences in January 2002 that the United States should ban such cloning for at least five years.

Past discussions of this subject have often given the impression that the safety concern is a purely temporary one that can be allayed in the near future, as scientific advances and improvements in technique reduce the risks to an ethically acceptable level. But this impression is mistaken, for considerable safety risks are likely to be enduring, perhaps permanent. If so, there will be abiding ethical difficulties *even with efforts aimed at making human cloning safe.*

The reason is clear: experiments to develop new reproductive technologies are necessarily intergenerational, undertaken to serve the reproductive desires of prospective parents but practiced also and always upon prospective children. Any such experiment unavoidably involves risks to the child-to-be, a being who is both the *product* and also the most vulnerable human *subject* of the research. Exposed to risk during the extremely sensitive life-shaping processes of his or her embryological development, any child-to-be is a singularly vulnerable creature, one maximally deserving of protection against risk of experimental (and other) harm. If experiments to learn how to clone a child are ever to be ethical, the degree of risk to that child-to-be would have to be extremely low, arguably no greater than for children-to-be who are conceived from union of egg and sperm. It is extremely unlikely that this moral burden can be met, not for decades if at all.

In multiple experiments involving six of the mammalian species cloned to date, more than 89 percent of the cloned embryos transferred to recipient females did not come to birth, and many of the live-born cloned animals are or become abnormal. If success means achieving normal and healthy development not just at birth but throughout the life span, there is even less reason for confidence. The oldest cloned mammal (Dolly) is only six years old and has exhibited unusually early arthritis. The reasons for failure in animal cloning are not well understood. Also, no nonhuman primates have been cloned. It will be decades (at least) before we could obtain positive evidence that cloned primates might live a normal healthy (primate) life.

Even a high success rate in animals would not suffice by itself to make human trials morally acceptable. In addition to the usual uncertainties in jumping the gap from animal to human research, cloning is likely to present particularly difficult problems of interspecies difference. Animal experiments have already shown substantial differences in the reproductive success of identical cloning techniques used in different species. If these results represent species-specific differences in, for example, the ease of epigenetic reprogramming and imprinting of the donor DNA, the magnitude of the risks to the child-to-be of the first human cloning experiments would be unknown and potentially large, no matter how much success had been achieved in animals. There can in principle be no direct experimental evidence sufficient for assessing the degree of such risk.

Can a highly reduced risk of deformity, disease, and premature death in animal cloning, coupled with the inherently unpredictable risk of moving from animals to humans, ever be low enough to meet the ethically acceptable standard set by reproduction begun with egg and sperm? The answer, as a matter of necessity, can never be better than "Just possibly." Given the severity of the possible harms involved in human cloning, and given that those harms fall on the very vulnerable child-to-be, such an answer would seem to be enduringly inadequate.

Similar arguments, it is worth noting, were made before the first attempts at human in vitro fertilization. People suggested that it would be unethical experimentation even to try to determine whether IVF could be safely done. And then, of course, IVF was accomplished. Eventually, it became a common procedure, and today the moral argument about its safety seems to many people beside the point. Yet the fact of success in that case does not establish precedent in this one, nor does it mean that the first attempts at IVF were not in fact unethical experiments upon the unborn, despite the fortunate results.

Be this as it may, the case of cloning is genuinely different. With IVF, assisted fertilization of egg by sperm immediately releases a developmental process, linked to the sexual union of the two gametes, that nature has selected over millions of years for the entire mammalian line. But in cloning experiments to produce children, researchers would be transforming a sexual system into an asexual one, a change that requires major and "unnatural" reprogramming of donor DNA if there is to be any chance of success. They are neither enabling nor restoring a natural process, and the alterations involved are such that success in one species cannot be presumed to predict success in another. Moreover, any new somatic mutations in the donor cell's chromosomal DNA would be passed along to the cloned child-to-be and its offspring. Here we can see even more the truly intergenerational character of cloning experimentation, and this should justify placing the highest moral burden of persuasion on those who would like to proceed with efforts to make cloning safe for producing children. (By reminding us of the need to protect the lives and well-being of our children and our children's children, this broader analysis of the safety question points toward larger moral objections to producing cloned children, objections that we shall consider shortly.)

It therefore appears to us that, given the dangers involved and the relatively limited goods to be gained from cloning-to-produce-children, conducting experiments in an effort to make cloning-to-produce-children safer would itself be an unacceptable violation of the norms of the ethics of research. *There seems to be no ethical way to try to discover whether cloning-to-produce-children can become safe, now or in the future.*

A Special Problem of Consent

A further concern relating to the ethics of human research revolves around the question of consent. Consent from the cloned child-to-be is of course impossible to obtain, and because no one consents to his or her own birth, it may be argued that concerns about consent are misplaced when applied to the unborn. But the issue is not so simple. For reasons having to do both with the safety concerns raised above and with the social, psychological, and moral concerns to be addressed below, an attempt to clone a human being would potentially expose a cloned individual-to-be to great risks of harm, quite distinct from those accompanying other sorts of reproduction. Given the risks, and the fact that consent cannot be obtained, the ethically correct choice may be to avoid the experiment.

The fact that those engaged in cloning cannot ask an unconceived child for permission places a burden on the cloners, not on the child. Given that anyone considering creating a cloned child must know that he or she is putting a newly created human life at exceptional risk, the burden on the would-be cloners seems clear: they must make a compelling case why the procedure should not be avoided altogether.

Reflections on the purpose and meaning of seeking consent support this point. Why, after all, does society insist upon consent as an essential principle of the ethics of scientific research? Along with honoring the free will of the subject, we insist on consent to protect the weak and the vulnerable, and in particular to protect them from the powerful. It would therefore be morally questionable, at the very least, to choose to impose potentially grave harm on an individual, especially in the very act of giving that individual life. Giving existence to a human being does not grant one the right to maim or harm that human being in research.

Problems of Exploitation of Women and Just Distribution of Risk

Cloning-to-produce-children may also lead to the exploitation of women who would be called upon to donate oocytes. Widespread use of the techniques of cloning-to-produce-children would require large numbers of eggs. Animal models suggest that several hundred eggs may be required before one attempt at cloning can be successful. The required oocytes would have to be donated, and the process of making them available would involve hormonal treatments to induce superovulation. If financial incentives are offered, they might lead poor women especially to place themselves at risk in this way (and might also compromise the voluntariness of their "choice" to make donations). Thus, research on cloning-to-produce-children could impose disproportionate burdens on women, particularly low-income women.

CONCLUSION

These questions of the ethics of research – particularly the issue of physical safety – point clearly to the conclusion that cloning-to-produce-children is unacceptable. In reaching this conclusion, we join the National Bioethics Advisory Commission and the National Academy of Sciences. But we go beyond the findings of those distinguished bodies in also pointing to the dangers that will *always* be inherent in the very process of trying to make cloning-to-produce-children safer. On this ground, we conclude that the problem of safety is not a temporary ethical concern. It is rather an enduring moral concern that might not be surmountable and should thus preclude work toward the development of cloning techniques to produce children. In light of the risks and other ethical concerns raised by this form of human experimentation, *we therefore conclude that cloning-to-produce-children should not be attempted.*

For some people, the discussion of ethical objections to cloning-to-produce-children could end here. Our society's established codes and practices in regard to human experimentation by themselves offer compelling reasons to oppose indefinitely attempts to produce a human child by cloning. But there *is* more to be said.

First, many people who are repelled by or opposed to the prospect of cloning human beings are concerned not simply or primarily because the procedure is unsafe. To the contrary, their objection is to the use of a *perfected* cloning technology and to a society that would embrace or permit the production of cloned children.

The ethical objection based on lack of safety is not really an objection to cloning *as such*. Indeed, it may in time become a vanishing objection should people be allowed to proceed – despite insuperable ethical objections such as the ones we have just offered– with experiments to perfect the technique. Should this occur, the ethical assessment of cloning-to-produce-children would need to address itself to the merits (and demerits) of cloning itself, beyond the safety questions tied to the techniques used to produce cloned children. Thus, anticipating the possibility of a perfected and usable technology, it is important to delineate the case against the practice itself.

Moreover, because the Council is considering cloning within a broad context of present and projected techniques that can affect human procreation or alter the genetic makeup of our children, it is important that we consider the full range and depth of ethical issues raised by such efforts.

How should these issues be raised, and within what moral framework? Some, but by no means all, of the deepest moral concerns connected to human cloning could be handled by developing a richer consideration of the ethics of human experimentation. Usually – and regrettably – we apply the ethical principles governing research on human subjects in a utilitarian spirit, weighing benefits versus harms, and moreover using only a very narrow notion of "harm." The calculus that weighs benefits versus harms too often takes stock only of bodily harm or violations of patient autonomy, though some serious efforts have been made in recent years to consider broader issues. In addition, we often hold a rather narrow view of what constitutes "an experiment." Yet cloning-to-produce-children would be a "human experiment" in many senses, and risks of bodily harm and inadequate consent do not exhaust the ways in which cloning might do damage. As we have described, cloning-to-produce-children would be a *biological experiment* – with necessary uncertainties about the safety of the technique and the possibility of physical harm. But it would also be an *experiment in human procreation* – substituting asexual for sexual reproduction and treating children not as gifts but as our self-designed products. It would be an *experiment in human identity* – creating the first human beings to inherit a genetic identity lived in advance by another. It would be an *experiment in genetic choice and design* – producing the first children whose entire genetic makeup was selected in advance. It would be an *experiment in family and social life* – altering the relationships within the family and between the generations, for example, by turning "mothers" into "twin sisters" and "grandparents" into "parents," and by having children asymmetrically linked biologically to only one parent. And it would represent a *social experiment* for the entire society, insofar as the society accepted, even if only as a minority practice, this unprecedented and novel mode of producing our offspring.

By considering these other ways in which cloning would constitute an experiment, we could enlarge our analysis of the ethics of research with human subjects to assess possible *nonbodily* harms of cloning-to-produce-children. But valuable as this effort might be, we have not chosen to proceed in this way. Not all the important issues can be squeezed into the categories of harms and benefits. People can be mistreated or done an injustice whether they know it or not and quite apart from any experienced harm. Important human goods can be traduced, violated, or sacrificed without being registered in anyone's catalogue of harms. The form of bioethical inquiry we are attempting here will make every effort not to truncate the moral meaning of our actions and practices by placing them on the Procrustean bed of utilitarianism. To be sure, the ethical principles governing human research are highly useful in efforts to protect vulnerable individuals against the misconduct or indifference of the powerful. But a different frame of reference is needed to evaluate the human meaning of innovations that may affect the lives and humanity of everyone, vulnerable or not.

Of the arguments developed below, some are supported by most Council Members, while other arguments are shared by only some Members. Even among the arguments they share, different Members find different concerns to be weightier. Yet we all believe that the arguments presented in the sections that follow are worthy of consideration in the course of trying to assess *fully* the ethical issues involved. We have chosen to err on the side of inclusion rather than exclusion of arguments because we acknowledge that concerns now expressed by only a few may turn out in the future to be more important than those now shared by all. Our fuller assessment begins with an attempt to fathom the deepest meaning of human procreation and thus necessarily the meaning of raising children. Our analysis will then move onto questions dealing with the effects of cloning on individuals, family life, and society more generally.

The Human Context: Procreation and Child-Rearing

Were it to take place, cloning-to-produce-children would represent a challenge to the nature of human procreation and child-rearing. Cloning is, of course, not only a means of procreation.

It is also a technology, a human experiment, and an exercise of freedom, among other things. But cloning would be most unusual, consequential, and most morally important as a new way of bringing children into the world and a new way of viewing their moral significance.

In Chapter One we outlined some morally significant features of human procreation and raised questions about how these would be altered by human cloning. We will now attempt to deepen that analysis, and begin with the salient fact that a child *is not made, but begotten*. Procreation is not making but the outgrowth of doing. A man and woman give themselves in love to each other, setting their projects aside in order to do just that. Yet a child results, arriving on its own, mysterious, independent, yet the fruit of the embrace. Even were the child wished for, and consciously so, he or she is the issue of their love, not the product of their wills; the man and woman in no way produce or choose a *particular* child, as they might buy a particular car. Procreation can, of course, be assisted by human ingenuity (as with IVF). In such cases, it may become harder to see the child solely as a gift bestowed upon the parents' mutual self-giving and not to some degree as a product of their parental wills. Nonetheless, because it is still sexual reproduction, the children born with the help of IVF begin – as do all other children – with a certain genetic independence of their parents. They replicate neither their fathers nor their mothers, and this is a salutary reminder to parents of the independence they must one day grant their children and for which it is their duty to prepare them.

Gifts and blessings we learn to accept as gratefully as we can. Products of our wills we try to shape in accord with our desires. Procreation as traditionally understood invites acceptance, rather than reshaping, engineering, or designing the next generation. It invites us to accept limits to our control over the next generation. It invites us even – to put the point most strongly – to think of the child as one who is not simply our own, our possession. Certainly, it invites us to remember that the child does not exist simply for the happiness or fulfillment of the parents.

To be sure, parents do and must try to form and mold their children in various ways as they inure them to the demands of family life, prepare them for adulthood, and initiate them into the human community. But, even then, it is only our sense that these children are not our possessions that makes such parental nurture which always threatens not to nourish but to stifle the child – safe.

This concern can be expressed not only in language about the relation between the generations but also in the language of equality. The things we make are not just like ourselves; they are the products of our wills, and their point and purpose are ours to determine. But a begotten child comes into the world just as its parents once did, and is therefore their equal in dignity and humanity.

The character of sexual procreation shapes the lives of children as well as parents. By giving rise to genetically new individuals, sexual reproduction imbues all human beings with a sense of individual identity and of occupying a place in this world that has never belonged to another. Our novel genetic identity symbolizes and foreshadows the unique, never-to-be-repeated character of each human life. At the same time, our emergence from the union of two individuals, themselves conceived and generated as we were, locates us immediately in a network of relation and natural affection.

Social identity, like genetic identity, is in significant measure tied to these biological facts. Societies around the world have structured social and economic responsibilities around the relationship between the generations established through sexual procreation, and have developed modes of child-rearing, family responsibility, and kinship behavior that revolve around the natural facts of begetting.

There is much more to be said about these matters, and they are vastly more complicated than we have indicated. There are, in addition, cultural differences in the way societies around the world regard the human significance of procreation or the way children are to be regarded and cared for. Yet we have said enough to indicate that the character and nature of human procreation matter deeply. They affect human life in endless subtle ways, and they shape families and communities. A proper regard for the profundity of human procreation (including child-rearing and parent-child relations) is, in our view, indispensable for a full assessment of the ethical implications of cloning-to-produce-children.

Identity, Manufacture, Eugenics, Family, and Society

Beyond the matter of procreation itself, we think it important to examine the possible psychological and emotional state of individuals produced by cloning, the well-being of their families, and the likely effects on society of permitting human cloning. These concerns would apply even if cloning-to-produce-children were conducted on a small scale; and they would apply in even the more innocent-seeming cloning scenarios, such as efforts to overcome infertility or to avoid the risk of genetic disease. Admittedly, these matters are necessarily speculative, for empirical evidence is lacking. Nevertheless, the importance of the various goods at stake justifies trying to think matters through in advance.

Keeping in mind our general observations about procreation, we proceed to examine a series of specific ethical issues and objections to cloning human children: (1) problems of identity and individuality; (2) concerns regarding manufacture; (3) the prospect of a new eugenics; (4) troubled family relations; and (5) effects on society.

Problems of Identity and Individuality

Cloning-to-produce-children could create serious problems of identity and individuality. This would be especially true if it were used to produce multiple "copies" of any single individual, as in one or another of the seemingly far-fetched futuristic scenarios in which cloning is often presented to the popular imagination. Yet questions of identity and individuality could arise even in small-scale cloning, even in the (supposedly) most innocent of cases, such as the production of a single cloned child within an intact family. Personal identity is, we would emphasize, a complex and subtle psychological phenomenon, shaped ultimately by the interaction of many diverse factors. But it does seem reasonably clear that cloning would at the very least present a unique and possibly disabling challenge to the formation of individual identity.

Cloned children may experience concerns about their distinctive identity not only because each will be genetically essentially identical to another human being, but also because they may resemble in appearance younger versions of the person who is their "father" or "mother." Of course, our genetic makeup does not by itself determine our identities. But our genetic uniqueness is an important source of our sense of who we are and how we regard ourselves. It is an emblem of independence and individuality. It endows us with a sense of life as a never-before-enacted possibility. Knowing and feeling that nobody has previously possessed our particular gift of natural characteristics, we go forward as genetically unique individuals into relatively indeterminate futures.

These new and unique genetic identities are rooted in the natural procreative process. A cloned child, by contrast, is at risk of living out a life overshadowed in important ways by the life of the "original" – general appearance being only the most obvious. Indeed, one of the reasons some people are interested in cloning is that the technique promises to produce in each case a particular individual whose traits and characteristics are already known. And however much or little one's genotype *actually* shapes one's natural capacities, it could mean a great deal to an individual's *experience* of life and the expectations that those who cloned him or her might have. The cloned child may be constantly compared to "the original," and may consciously or unconsciously hold himself or herself up to the genetic twin that came before. If the two individuals turned out to lead similar lives, the cloned person's achievements may be seen as derivative. If, as is perhaps more likely, the cloned person departed from the life of his or her progenitor, this very fact could be a source of constant scrutiny, especially in circumstances in which parents produced their cloned child to become something in particular. Living up to parental hopes and expectations is frequently a burden for children; it could be a far greater burden for a cloned individual. The shadow of the cloned child's "original" might be hard for the child to escape, as would parental attitudes that sought in the child's very existence to replicate, imitate, or replace the "original."

It may reasonably be argued that genetic individuality is not an indispensable human good, since identical twins share a common genotype and seem not to be harmed by it. But this argument misses the context and environment into which even a single human clone would be born. Identical twins have as progenitors two biological parents and are born together, before either one has developed and shown what his or her potential – natural or otherwise – may be. Each is largely free of the burden of measuring up to or even knowing in advance the genetic traits of the other, because both begin life together and neither is yet known to the world. But a clone is a genetic near-copy of a person who is already living or has already lived. This might constrain the clone's sense of self in ways that differ in kind from the experience of identical twins. Everything about the predecessor – from physical height and facial appearance, balding patterns and inherited diseases, to temperament and native talents, to shape of life and length of days, and even cause of death – will appear before the expectant eyes of the cloned person, always with at least the nagging concern that there, notwithstanding the grace of God, go I. The crucial matter, again, is not simply the truth regarding the extent to which genetic identity actually shapes us – though it surely does shape us to some extent. What matters is the cloned individual's *perception* of the significance of the "precedent life" and the way that perception cramps and limits a sense of self and independence.

Concerns regarding Manufacture

The likely impact of cloning on identity suggests an additional moral and social concern: the transformation of human procreation into human manufacture, of begetting into making. By using the terms "making" and "manufacture" we are not claiming that cloned children would be artifacts made altogether "by hand" or produced in factories. Rather, we are suggesting that they would, like other human "products," be brought into being in accordance with some pre-selected genetic pattern or design, and therefore in some sense "made to order" by their producers or progenitors.

Unlike natural procreation – or even most forms of assisted reproduction – cloning-to-produce-children would set out to create a child with a very particular genotype: namely, that of the somatic cell donor. Cloned children would thus be the first human beings whose entire genetic makeup is selected in advance. True, selection from among existing genotypes is not yet design of new ones. But the principle that would be established by human cloning is both far-reaching and completely novel: parents, with the help of science and technology, may determine in advance the genetic endowment of their children. To this point, parents have the right and the power to decide *whether* to have a child. With cloning, parents acquire the power, and presumably the right, to decide *what kind* of a child to have. Cloning would thus extend the power of one generation over the next – and the power of parents over their offspring – in ways that open the door, unintentionally or not, to a future project of genetic manipulation and genetic control.

Of course, there is no denying that we have already taken steps in the direction of such control. Preimplantation genetic diagnosis of embryos and prenatal diagnosis of fetuses – both now used to prevent the birth of individuals carrying genes for genetic diseases – reflect an only conditional acceptance of the next generation. With regard to *positive* selection for desired traits, some people already engage in the practice of sex selection, another example of conditional acceptance of offspring. But these precedents pale in comparison to the degree of control provided by cloning and, in any case, do not thereby provide a license to proceed with cloning. It is far from clear that it would be wise to proceed still farther in our attempts at control.

The problem with cloning-to-produce-children is not that artificial technique is used to assist reproduction. Neither is it that genes are being manipulated. We raise no objection to the use of the coming genetic technologies to treat individuals with genetic diseases, even in utero – though there would be issues regarding the protection of human subjects in research and the need to find boundaries between therapy and so-called enhancement (of this, more below). The problem has to do with the control of the entire genotype and the production of children to selected specifications.

Why does this matter? It matters because human dignity is at stake. In natural procreation, two individuals give life to a new human being whose endowments are not shaped deliberately by human will, whose being remains mysterious, and the open-endedness of whose future is ratified and embraced. Parents beget a child who enters the world exactly as they did – as an unmade gift, not as a product. Children born of this process stand equally beside their progenitors as fellow human beings, not beneath them as made objects. In this way, the uncontrolled beginnings of human procreation endow each new generation and each new individual with the dignity and freedom enjoyed by all who came before.

Most present forms of assisted reproduction imitate this natural process. While they do begin to introduce characteristics of manufacture and industrial technique, placing nascent human life for the first time in human hands, they do not control the final outcome. The end served by IVF is still the same as natural reproduction-the birth of a child from the union of gametes from two progenitors. Reproduction with the aid of such techniques still implicitly expresses a willingness to accept as a gift the product of a process we do not control. In IVF children emerge out of the same mysterious process from which their parents came, and are therefore not mere creatures of their parents.

By contrast, cloning-to-produce-children – and the forms of human manufacture it might make more possible in the future – seems quite different. Here, the process begins with a very specific final product in mind and would be tailored to produce that product. Even were cloning to be used solely to remedy infertility, the decision to clone the (sterile) father would be a decision, willy-nilly, that the child-to-be should be the near-twin of his "father."

Anyone who would clone merely to ensure a "biologically related child" would be dictating a very specific form of biological relation: genetic virtual identity. In every case of cloning-to-produce-children, scientists or parents would set out to produce specific individuals for particular reasons. The procreative process could come to be seen increasingly as a means of meeting specific ends, and the resulting children would be products of a designed manufacturing process, products over whom we might think it proper to exercise "quality control." Even if, in any given case, we were to continue to think of the cloned child as a gift, *the act itself teaches a different lesson*, as the child becomes the continuation of a parental project. We would learn to receive the next generation less with gratitude and surprise than with control and mastery.

One possible result would be the industrialization and commercialization of human reproduction. Manufactured objects become commodities in the marketplace, and their manufacture comes to be guided by market principles and financial concerns. When the "products" are human beings, the "market" could become a profoundly dehumanizing force. Already there is commerce in egg donation for IVF, with ads offering large sums of money for egg donors with high SAT scores and particular physical features.

The concerns expressed here do not depend on cloning becoming a widespread practice. The introduction of the terms and ideas of production into the realm of human procreation would be troubling regardless of the scale involved; and the adoption of a market mentality in these matters could blind us to the deep moral character of bringing forth new life. Even were cloning children to be rare, the moral harms to a society that accepted it could be serious.

Prospect of a New Eugenics

For some of us, cloning-to-produce-children also raises concerns about the prospect of eugenics or, more modestly, about genetic "enhancement." We recognize that the term "eugenics" generally refers to attempts to improve the genetic constitution of a particular political community or of the human race through general policies such as population control, forced sterilization, directed mating, or the like. It does not ordinarily refer to actions of particular individuals attempting to improve the genetic endowment of their own descendants. Yet, although cloning does not in itself point to public policies by which the state would become involved in directing the development of the human gene pool, this might happen in illiberal regimes, like China, where the government already regulates procreation. And, in liberal societies, cloning-to-produce-children could come to be used privately for individualized eugenic or "enhancement" purposes: in attempts to alter (with the aim of improving) the genetic constitution of one's own descendants – and, indirectly, of future generations.

Some people, in fact, see enhancement as the major purpose of cloning-to-produce-children. Those who favor eugenics and genetic enhancement were once far more open regarding their intentions to enable future generations to enjoy more advantageous genotypes. Toward these ends, they promoted the benefits of cloning: escape from the uncertain lottery of sex, controlled and humanly directed reproduction. In the present debate about cloning-to-produce-children, the case for eugenics and enhancement is not made openly, but it nonetheless remains an important motivation for some advocates. Should cloning-to-produce-children be introduced successfully, and should it turn out that the cloned humans do in fact inherit many of the natural talents of the "originals," some people may become interested in the prospects of using it to produce "enhanced children" – especially if other people's children were receiving comparable advantages.

Cloning can serve the ends of individualized enhancement either by avoiding the genetic defects that may arise when human reproduction is left to chance or by preserving and perpetuating outstanding genetic traits. In the future, if techniques of genetic enhancement through more precise genetic engineering became available, cloning could be useful for perpetuating the enhanced traits and for keeping any "superior" manmade genotype free of the flaws that sexual reproduction might otherwise introduce.

"Private eugenics" does not carry with it the dark implications of state despotism or political control of the gene pool that characterized earlier eugenic proposals and the racist eugenic practices of the twentieth century. Nonetheless, it could prove dangerous to our humanity. Besides the dehumanizing prospects of the turn toward manufacture that such programs of enhancement would require, there is the further difficulty of the lack of standards to guide the choices for "improvement." To this point, biomedical technology has been applied to treating diseases in patients and has been governed, on the whole, by a commonsense view of health and disease. To be sure, there are differing views about how to define "health." And certain cosmetic, performance – enhancing, or hedonistic uses of biomedical techniques have already crossed any plausible boundary between therapy and enhancement, between healing the sick and "improving" our powers. Yet, for the most part, it is by some commonsense views of health that we judge who is in need of medical treatment and what sort of treatment might be most appropriate. Even today's practice of a kind of "negative" eugenics–through prenatal genetic diagnosis and abortion of fetuses with certain genetic abnormalities – is informed by the desire to promote health.

The "positive" eugenics that could receive a great boost from human cloning, especially were it to be coupled with techniques of precise genetic modification, would not seek to restore sick human beings to natural health. Instead, it would seek to alter humanity, based upon subjective or arbitrary ideas of excellence. The effort may be guided by apparently good intentions: to improve the next generation and to enhance the quality of life of our descendants. But in the process of altering human nature, we would be abandoning the standard by which to judge the goodness or the wisdom of the particular aims. We would stand to lose the sense of what is and is not human.

The fear of a new eugenics is not, as is sometimes alleged, a concern born of some irrational fear of the future or the unknown. Neither is it born of hostility to technology or nostalgia for some premodern pseudo-golden age of superior naturalness. It is rather born of the rational recognition that once we move beyond therapy into efforts at enhancement, we are in uncharted waters without a map, without a compass, and without a clear destination that can tell us whether we are making improvements or the reverse. The time-honored and time-tested goods of human life, which we know to be good, would be put in jeopardy for the alleged and unknowable goods of a post-human future.

Troubled Family Relations

Cloning-to-produce-children could also prove damaging to family relations, despite the best of intentions. We do not assume that cloned children, once produced, would not be accepted, loved, or nurtured by their parents and relatives. On the contrary, we freely admit that, like any child, they might be welcomed into the cloning family. Nevertheless, the cloned child's place in the scheme of family relations might well be uncertain and confused. The usually clear designations of father and brother, mother and sister, would be confounded. A mother could give birth to her own genetic twin, and a father could be genetically virtually identical to his son. The cloned child's relation to his or her grandparents would span one and two generations at once. Every other family relation would be similarly confused. There is, of course, the valid counter-argument that holds that the "mother" could easily be defined as the person who gives birth to the child, regardless of the child's genetic origins, and for social purposes that may serve to eliminate some problems. But because of the special nature of cloning-to-produce-children, difficulties may be expected.

The crucial point is not the absence of the natural biological connections between parents and children. The crucial point is, on the contrary, the presence of a unique, one-sided, and replicative biological connection to only one progenitor. As a result, family relations involving cloning would differ from all existing family arrangements, including those formed through adoption or with the aid of IVF. A great many children, after all, are adopted, and live happy lives in loving families, in the absence of any biological connections with their parents. Children conceived by artificial insemination using donor sperm and by various IVF techniques may have unusual relationships with their genetic parents, or no genetic relationships at all. But all of these existing arrangements attempt in important ways to emulate the model of the natural family (at least in its arrangement of the generations), while cloning runs contrary to that model.

What the exact effects of cloning-to-produce-children might be for families is highly speculative, to be sure, but it is still worth flagging certain troubling possibilities and risks. The fact that the cloned child bears a special tie to only one parent may complicate family dynamics. As the child developed, it could not help but be regarded as specially akin to only one of his or her parents. The sins or failings of the father (or mother), if reappearing in the cloned child, might be blamed on the progenitor, adding to the chances of domestic turmoil. The problems of being and rearing an adolescent could become complicated should the teenage clone of the mother "reappear" as the double of the woman the father once fell in love with. Risks of competition, rivalry, jealousy, and parental tension could become heightened.

Even if the child were cloned from someone who is not a member of the family in which the child is raised, the fact would remain that he or she has been produced in the nearly precise genetic image of another and for some particular reason, with some particular design in mind. Should this become known to the child, as most likely it would, a desire to seek out connection to the "original" could complicate his or her relation to the rearing family, as would living consciously "under the reason" for this extra-familial choice of progenitor. Though many people make light of the importance of biological kinship (compared to the bonds formed through rearing and experienced family life), many adopted children and children conceived by artificial insemination or IVF using donor sperm show by their actions that they do not agree. They make great efforts to locate their "biological parents," even where paternity consists in nothing more than the donation of sperm. Where the progenitor is a genetic near-twin, surely the urge of the cloned child to connect with the unknown "parent" would be still greater.

For all these reasons, the cloning family differs from the "natural family" or the "adoptive family." By breaking through the natural boundaries between generations, cloning could strain the social ties between them.

Effects on Society

The hazards and costs of cloning-to-produce-children may not be confined to the direct participants. The rest of society may also be at risk. The impact of human cloning on society at large may be the least appreciated, but among the most important, factors to consider in contemplating the morality of this activity.

Cloning is a human activity affecting not only those who are cloned or those who are clones, but also the entire society that allows or supports such activity. For insofar as the society accepts cloning-to-produce-children, to that extent the society may be said to engage in it. A society that allows dehumanizing practices – especially when given an opportunity to try to prevent them – risks becoming an accomplice in those practices. (The same could be said of a society that allowed even a few of its members to practice incest or polygamy.) Thus the question before us is whether cloning-to-produce-children is an activity that we, as a society, should engage in. In addressing this question, we must reach well beyond the rights of individuals and the difficulties or benefits that cloned children or their families might encounter. We must consider what kind of a society we wish to be, and, in particular, what forms of bringing children into the world we want to encourage and what sorts of relations between the generations we want to preserve.

Cloning-to-produce-children could distort the way we raise and view children, by carrying to full expression many regrettable tendencies already present in our culture. We are already liable to regard children largely as vehicles for our own fulfillment and ambitions. The impulse to create "designer children" is present today – as temptation and social practice. The notion of life as a gift, mysterious and limited, is under siege. Cloning-to-produce-children would carry these tendencies and temptations to an extreme expression. It advances the notion that the child is but an object of our sovereign mastery.

A society that clones human beings thinks about human beings (and especially children) differently than does a society that refuses to do so. It could easily be argued that we have already in myriad ways begun to show signs of regarding our children as projects on which we may work our wills. Further, it could be argued that we have been so desensitized by our earlier steps in this direction that we do not recognize this tendency as a corruption.

While some people contend that cloning-to-produce-children would not take us much further down a path we have already been traveling, we would emphasize that the precedent of treating children as projects cuts two ways in the moral argument.

Instead of using this precedent to justify taking the next step of cloning, the next step might rather serve as a warning and a mirror in which we may discover reasons to reconsider what we are already doing. Precisely because the stakes are so high, precisely because the new biotechnologies touch not only our bodies and minds but also the very idea of our humanity, we should ask ourselves how we as a society want to approach questions of human dignity and flourishing.

CONCLUSION

Cloning-to-produce-children may represent a forerunner of what will be a growing number of capacities to intervene in and alter the human genetic endowment. No doubt, earlier human actions have produced changes in the human gene pool: to take only one example, the use of insulin to treat diabetics who otherwise would have died before reproducing has increased the genes for diabetes in the population. But different responsibilities accrue when one sets out to make such changes prospectively, directly, and deliberately. To do so without regard for the likelihood of serious unintended and unanticipated consequences would be the height of hubris. Systems of great complexity do not respond well to blunt human intervention, and one can hardly think of a more complex system – both natural and social – than that which surrounds human reproduction and the human genome. Given the enormous importance of what is at stake, we believe that the so-called "precautionary principle" should be our guide in this arena. This principle would suggest that scientists, technologists, and, indeed, all of us should be modest in claiming to understand the many possible consequences of any profound alteration of human procreation, especially where there are not compelling reasons to proceed. Lacking such understanding, no one should take action so drastic as the cloning of a human child. In the absence of the necessary human wisdom, prudence calls upon us to set limits on efforts to control and remake the character of human procreation and human life.

It is not only a matter of prudence. Cloning-to-produce-children would also be an injustice to the cloned child – from the imposition of the chromosomes of someone else, to the intentional deprivation of biological parents, to all of the possible bodily and psychological harms that we have enumerated in this chapter. It is ultimately the claim that the cloned child would be seriously wronged – and not only harmed in body – that would justify government intervention. It is to this question – the public policy question of what the government should and can do to prevent such injustice – that we will turn in Chapter Seven. But, regarding the ethical assessment, Members of the Council are in unanimous agreement that cloning-to-produce-children is not only unsafe but also morally unacceptable and ought not to be attempted.

12

The Ethics of Cloning-for-Biomedical-Research

THE MANNER AND SPIRIT OF THIS INQUIRY

The question of whether or not to proceed with human cloning-for-biomedical-research is a morally serious and difficult one. On the one hand, there is the promise that such research could lead to important knowledge of human embryological development and gene action, especially in cases in which there are genetic abnormalities that lead to disease. There is also the promise that such research could contribute to producing transplantable tissues and organs that could be effective in curing or reversing many dreaded illnesses and injuries – including Parkinson's disease, Alzheimer's disease, juvenile diabetes, and spinal cord injury. On the other hand, there are the morally relevant facts that this research involves the deliberate production, use, and ultimate destruction of cloned human embryos, and that the cloned embryos produced for research are no different from cloned embryos that could be used in attempts to produce cloned children.

Complicating the moral assessment are questions about the likelihood that this research will deliver its promised benefits and about the possibility of equally promising, yet morally less problematic, approaches to the same scientific and medical goals. Finally, there is the ever-present danger of creating false hope among patients, and the risk of allowing the goodness of the end (finding cures for disease) to justify moral indifference to the means used to achieve it. Morally serious people may differ in their final judgment of the ethics of cloning-for-biomedical-research. But they do – or should – agree on this: that fidelity both to the highest moral and human aspirations of science and medicine and to the moral standards of the wider community requires that we consider not only why and how to proceed with new lines of research, but also whether there might be compelling reasons not to do so or certain limits that should be observed. Both the facts (scientific and moral) and our ethical principles must be consulted in trying to judge what is best.

Yet despite this general agreement, it is difficult to know how best to proceed in the present case. There are multiple questions about the right context for considering the ethics of cloning-for-biomedical-research. First, we must weigh whether to take up this matter in the context of deciding what to do about cloning-to-produce-children or in the somewhat different context of the ethics of embryo and stem cell research more generally. The issue has in fact emerged in the public moral debate over anti-cloning legislation, as a complication in the effort to stop cloning-to-produce-children. Generally speaking, the most effective way to prevent cloning-to-produce-children would arguably be to stop the process at the initial act of cloning, the production (by an act of somatic cell nuclear transfer [SCNT]) of the embryonic human clone. Yet such a measure would rule out cloning-for-biomedical-research, and many scientists and patient advocacy groups have argued that the human and moral costs of doing so are too great. Alternatively, we could take up this matter in what seems philosophically to be its more natural context, namely, as a sub-species of a larger inquiry into the ethics of embryo and stem cell research.

Each of these contexts – what to do about cloning-to-produce-children and what to do about embryo research – is certainly plausible. Yet each, by itself, is less than satisfactory. The first risks giving excessive weight to the fact that the embryos wanted for research are *cloned* embryos; the second, ignoring the aspect (central to cloning) of *genetic manipulation*, risks the opposite error by requiring that the ethics of cloning-for-biomedical-research be argued entirely in terms of what it is proper to do with *embryos as such*. We can imagine, in advance of any discussion, a variety of moral opinions that would emerge, influenced in part by how the question is formulated: one person could defend stem cell research performed using embryos produced by IVF but oppose research using cloned embryos for reasons of prudence (such as decreasing the likelihood of cloning-to-produce-children). Another person, holding IVF embryos in higher regard than cloned embryos, could reach precisely the opposite conclusion. Some people will hold that research on any human embryo, cloned or not, is always morally unacceptable (or acceptable), independent of whether ethical or legal guidelines are in place, while others will judge one way or another depending on whether appropriate guidelines and effective regulations have been established.

We have decided to discuss the ethics of cloning-for-biomedical-research in the broader moral – philosophical context, rather than the narrower moral – political one that has taken shape around the current debate over anti-cloning legislation. Though we are mindful of the importance of these public policy debates – and will consider them in the following chapter – we do not want our moral analysis to be skewed by the specific legal or policy questions at issue, especially as the moral questions discussed here have implications beyond the current political debate and even beyond the question of human cloning itself. We opt to take up the moral questions in their fullness.

A second question about context is even more difficult to assess. Should we regard cloning-for-biomedical-research as just the latest – and continuous – step in trying to unlock the secrets of human development and to discover cures for diseases? Or should it be seen – instead or also – as the earliest stage of a revolutionary new science of enhancement or eugenics, which will go beyond treating individuals with disease and disability to attempt engineered improvements in human genetic endowments? Because innovations like cloning come to us gradually and piecemeal, and because it cannot be known in advance how exactly they will be used or where they might lead, there is a temptation to stay close to the present and to ignore possible future implications.

Yet the alleged perils of going ahead with the research are arguably no more speculative than the promised benefits. And it would be morally and prudently shortsighted for this Council, charged with investigating "the human and moral significance of advances in biomedical science and technology," to refuse to think about where this research might lead. We will therefore consider, even if we cannot know in advance, whether and how the sort of genetic manipulation of embryos exemplified by cloning-for-biomedical-research is new or "revolutionary." Genetic therapy for existing diseases and non-therapeutic genetic modifications of our native endowments raise profoundly different questions. Accordingly, we will keep in sight not only the moral questions surrounding the *means* of cloning-for-biomedical-research – which is to say, the significance of using or not using nascent human life as a resource – but also the possible *ends* to which our expanding knowledge and capacities might be put. At the same time, we will be careful not to equate genetic medicine that is truly in the service of human life with genetic manipulation that is not, and to avoid both the unjustified fear and exaggerated promises that sometimes accompany biomedical progress.

A third difficulty concerns the relation between the ethics of research on embryos (cloned or not) and the ethics of abortion. For many people, these issues are linked, and there is doubtless an overlap in the moral questions involved. Yet the issues are, in important respects, quite distinct. In the case of abortion, the fetuses whose fate is at issue are unwanted and (usually) the result of unintended conception.

The embryos produced for research are wanted, indeed deliberately created, with certain knowledge and intent that they will be used and destroyed. More important perhaps, the extra-corporeal embryo (whether produced specially for research or left-over in IVF procedures) does not exist in conflict with the wishes, interests, or rights of a woman who is pregnant.

Also, although abortion is widely practiced, each decision to abort is made one at a time, case by case. In contrast, to embark on creating cloned embryos only for purposes of research is to countenance at one stroke the large-scale production of developing human life for routinized use and destruction. For these reasons, we shall try to consider the question of the ethics of research on embryos in its own terms, distinct from the ethical questions about abortion.

Finally, there is the question of the spirit in which this examination should be conducted. Reflecting the situation in American society, there are major differences within the Council regarding the morality of research involving early human (cloned) embryos. These differences turn largely, though not exclusively, on different judgments regarding the nature and moral status of the early human (cloned) embryo: namely, to what extent is it, or is it not, "one of us," a human life in process? Having explored these questions collegially among ourselves, we have come to think that *all* parties to this debate have something vital to defend, something vital not only to themselves but *also to their opponents in the debate,* and indeed to all human beings. No human being and no human society can afford to be callous to the needs of suffering humanity, cavalier regarding the treatment of nascent human life, or indifferent to the social effects of adopting in these matters one course of action rather than another.

We believe, therefore, that we can make our best contribution to a truthful and appropriate moral understanding of the issue by developing, in a single document, the moral cases both *for* and *against* proceeding with cloning-for-biomedical-research (and also articulating, where necessary and as clearly as possible, important differences within each of these cases). Each Member of the Council has been asked to help strengthen the case made for both sides, regardless of which side he or she inclines toward. By proceeding in this way, we hope to make clear to the President and the nation exactly what is morally and humanly at stake in the controversy and what may be gained *and lost* in whatever choice is finally made.

Thus, notwithstanding our differences, we stand together as the authors of the entire chapter, hoping by this means to shed light rather than heat on this most vexing of moral and policy questions. At the same time, we have tried fully and fairly to articulate our differences, and to do so by speaking, in the first person, as members of a deliberative body called upon to make our own best judgments. *This means that the "we" that now embraces all Members of the Council will stand in the particular sections presenting the moral case for and the moral case against cloning-for-biomedical-research,* **only** *for those among us who subscribe to the specific arguments* *being made in those sections. In other words, each opinion is a self-contained brief, representing not the Council as a whole but only a portion of the Council. And even within the cases for and against, Members of the Council disagree over matters of substance and emphasis.* But while the Council has strong differences of opinion, as delineated in the sections that follow, the Council speaks in a single voice in its affirmation that the debate must not be won by dismissing, ridiculing, or demonizing the other side. Important human goods are to be found on all sides of the debate, a fact too often overlooked.

THE HUMAN MEANING OF HEALING

Before presenting the two opinions, we will place the moral questions surrounding cloning-for-biomedical-research in their larger human context. Just as we did in discussing the ethics of cloning-to-produce-children, we step back from the particular technological possibility at hand to look carefully at the larger human goods that we seek both to serve and defend. We look specifically at the human meaning of healing the sick and aiding the suffering, as well as the spirit and practice of biomedical research that aims to make such healing possible. This exploration will better prepare us to see what is humanly at stake in our moral judgment about cloning-for-biomedical-research, and to face soberly both what is gained and what is lost in either proceeding or not proceeding. The subsequent moral arguments, both pro and con, are informed by these larger reflections.

To be human is to be mortal. To be alive is to be vulnerable to suffering. No one is better situated to appreciate these truths than the physician. To understand what it means to heal, one must therefore understand the doctor's special encounter with human suffering – as both an experience (a crying out) of the patient who lies before him and as a central mystery of human existence. Why do human beings suffer? Why do they suffer in ways that cannot be explained – entirely or perhaps at all – with human notions of justice? In this role, the doctor is sometimes a *messenger* of human finitude. He must tell patients that their days are numbered or that their time has come; he must tell grieving family members that death is at the door. But the healer is also and more importantly – in the eyes of both doctor and patient – a *deliverer*. Not only is he well armed to deliver us from specific maladies and miseries. He is also a much needed ally against the deadly disease – traditionally regarded as a sin – of despair. Because of the moral aspirations of his calling, the physician is a trusted source of hope that the living might yet still live and that in his skill and the powerful techniques of modern medicine might lie the possibility of renewal. The doctor is, at different times, a reminder of the intractable sadness of human life, but also explicitly a conqueror who beats back suffering and disease with the saving hand of medical knowledge and technique, and who inspirits us with hope to go forward even in the absence of cure and relief.

Until roughly the second half of the twentieth century, physicians delivered more hope than cure, and they conquered few diseases. Since then, their arsenal against disease (at least in technologically advanced nations) has grown enormously, and it promises to grow greater in the decades ahead. New healing powers will surely emerge from the work of medicine's ally, biomedical research, firmly grounded in the principles and methods of modern biomedical science. This noble field of human endeavor also has a context in the larger domain of human life. Celebrating its achievements and eager for its gifts to human welfare, modern societies embrace and invest heavily in medical research and grant scientists great freedom to inquire and experiment. Because of the way science advances, freedom is crucial to the successful realization of its goals.

Dr. William Osler, one of the founding figures of modern medicine, described the aspirations of biomedical research as follows:

To wrest from nature the secrets which have perplexed philosophers in all ages, to track to their sources the causes of disease, to correlate the vast stores of knowledge that they may be quickly available for the prevention and cure of disease – These are our ambitions.

It is in the very nature of a "secret" that one cannot know in advance which areas of research and discovery will prove the most fruitful. One proceeds by trial and error. One makes hypotheses grounded in what is already known, in the effort to discover what remains a mystery. One begins with basic research into disease processes and mechanisms, in the hope that new knowledge will yield new medicines and new cures.

One motive for such research is simply the love of knowledge itself – the distinctively human desire to know, to see, to understand more than one already does. But biomedical research is also guided, above all, by the humanitarian desire to apply new knowledge in the service of those who suffer, to correlate knowledge that it "may be quickly available for the prevention and cure of disease." Biomedical scientists aim to weld the virtues of charity, beneficence, and responsibility to the human ambition to "wrest from nature" her secrets. This is the moral heart of both the medical profession and the research tradition that supports it: to do everything in our power, consistent with law and morals, to provide cures, amelioration, and relief to those who need them.

"Consistent with law and morals": this requirement qualifies "everything in our power." This limitation has been traditionally understood to be part of the healing vocation. Moral philosophers and philosophers of medicine have long held that the duty to heal is an "imperfect duty," meaning that it does not trump all other considerations. Physicians perhaps understand this best of all, learning their limits empirically from their encounters with patients whom they cannot save or even comfort. The duty to heal *this* patient, *at this time*, is also an imperfect one. After all, a cure for one person at the direct expense of another – for example, harvesting a vital organ from someone who is living to save someone else who is dying – would violate the first principle of medicine to "do no harm."

It is also true that scientific freedom and medical progress are not the only human goods worthy of our commitment and protection. Research must be judged both by the means it employs and by the ends it serves (both those that were intended and those that were not). The Nuremberg Code, the Helsinki Declaration, and the Belmont Report, discussed in the last chapter, are all efforts to set moral limits on biomedical research and to ensure that science serves human beings rather than the other way around. Among other things, these ethical codes embody the recognition that those who do research *about* human beings can never escape (nor should they) their status *as* human beings. Those who investigate human biology are always both the *knowers* and *the subject that* is known, both the potential healers and the potentially afflicted. And therefore they must never treat that which is their equal – their fellow human beings – as something less than human.

But in the end, however imperfect it is as a duty and whatever its less than supreme place among all other human goods, the obligation to heal and to seek remedies is a powerful one. It is a mark both of man's natural limits (as the being in need of healing) and his capacity for goodness (as the being who heals). And so, the freedom of inquiry that makes biomedical research possible should be restricted only for the most important reasons, lest we do damage to the entire enterprise, or to the human beings and the society that benefit from the "vast stores of knowledge" it creates.

At the same time, however, those who have accepted the "healer's covenant" – and those who defend, engage in, and benefit from the research that improves and expands the human capacity to heal – must avoid the seduction of medical triumphalism: the belief that all human suffering, both physical and psychic, can be conquered by modern technique, and therefore that no form of biomedical research should be opposed. Doctors and scientists must not become partial human beings who evade moral responsibility by claiming that they are not qualified to judge the moral implications of their own medical research or, worse, that medically beneficial research is always self-justifying, and hence that there are no real moral dilemmas at all. In addition, they must avoid the cruelty of creating false hopes among patients and their loved ones, and the folly of creating messianic or utopian visions of what science and medicine can accomplish. And patients, even as they heroically fight against suffering, must not forget their own mortality – including the often unpredictable nature of how and when death comes.

These reflections point to the following conclusions: In judging the moral beneficence and moral hazards of medical research, we must remember that suffering should not be opposed by any means possible. We would be less than human if we did not desire to alleviate such suffering, but we would be imagining ourselves to be more than human if we thought and acted as if we could alleviate it once and for all. Rather, we must acknowledge that as human beings we live in a difficult "in-between." Whether as doctors, scientists, or as patients, we all wish for the possible renewal of life through medicine, but also acknowledge that suffering and mortality are part of being alive. We are morally obliged to seek relief of suffering, but only in ways that preserve our humanity.

With these realities in mind, this chapter will now take up the ethics of cloning-for-biomedical-research, and specifically the moral and human questions raised above: What is owed to those who suffer from debilitating injuries and diseases? What is owed to nascent human life? And what is owed to the moral well-being of society? These are the central questions in the debate, questions that Members of the Council over the past year struggled to answer, and that indeed every member of society must ponder when considering the ethics of cloning-for-biomedical-research.

The moral case for cloning-for-biomedical-research can be stated in the following straightforward way: American society and human communities in general have an obligation to try to heal the sick and relieve their suffering. This obligation, deeply rooted in the moral teaching of "love of neighbor," lies heaviest on physicians and health-care professionals who attend to individual patients. But it guides also the activities of biomedical scientists and biotechnologists whose pioneering research and discoveries provide new and better means of healing and relieving those who suffer. Research on cloned human embryos is one more path to discovering such means. Like embryonic stem cell research, to which it is partially related, it offers a promising approach to gaining knowledge and techniques that could lead to new treatments for chronic genetic or acquired degenerative diseases and disabilities. If successful, it could help save countless human lives and ameliorate untold human suffering.

It is true that human cloning-for-biomedical-research raises ethical questions, mainly because it involves the production, use, and destruction of cloned human embryos. It is also true that cloned embryos produced for research could be used in attempts to produce cloned human children, and the availability of such cloned embryos for research and the perfection of cloning techniques might increase the likelihood that people will succeed in cloning children. We appreciate the concerns of people who voice these objections and risks, and we are prepared to accept certain limits and safeguards against possible abuses. Yet we believe that, on balance, the objections to cloning-for-biomedical-research are outweighed by the good that can be done for current and future individuals who suffer. The moral balance lies on the side of endorsing and encouraging this activity.

We who endorse cloning-for-biomedical-research will attempt to make a version of this case here. But we will do so, for the most part, in a somewhat different spirit, one that is informed by the discussion of healing just concluded. In moral debates about these matters, people often speak as if saving lives is the only value that counts and that everything else must be sacrificed to advancing potentially beneficial research. Others speak as if any failure to prevent death or suffering from disease is sinful. Our defense of cloning-for-biomedical-research is more complex and nuanced and, we believe, more true to the merits of the case in question. As we make our case, we will also confront-and accept-the burden of what it means to proceed with such research, just as those who oppose it must accept the burden of what it means not to proceed.

In making our case, we begin in Section A by summarizing the specific medical benefits that might be achieved by proceeding with this avenue of research. We then consider in Section B the moral dilemmas of this research. However, among those of us who believe the research should go forward there is disagreement about how seriously to take certain moral objections, and thus two distinct positions for proceeding are presented.

The Medical Promise of Cloning-for-Biomedical-Research

Many people suffer from chronic debilitating diseases and disabilities, including, among others, juvenile diabetes, Parkinson's disease, Alzheimer's disease, spinal cord injuries, heart disease, and amyotrophic lateral sclerosis. These terrible diseases shorten life, limit activity (often severely), and cause great suffering both for the afflicted and their families. The inspiring example of exceptional persons who bear bravely the great burdens of illness or injury should not blind us to the powerful warrants for research and therapy that might lift these burdens. The likelihood of premature death, in particular, can shadow the life of the patient and the patient's family even before it arrives, and its advent can impoverish and devastate families, dash hopes, and cast a chill on the lives of survivors. It is certainly admirable to confront, endure, and redeem these unchosen afflictions. But it is also admirable, where possible, to ameliorate through research and medicine the diseases and injuries that cause them.

Cloning-for-biomedical-research may offer unique ways of investigating and possibly treating several of these diseases. To unlock the secrets of a disease, scientists must explore its specific molecular and cellular mechanisms, carefully observing both normal and pathological development. This research could be greatly facilitated by in vitro cellular models of human disease. It is here that the potentially most valuable and unique benefits of research on cloned human embryos may lie. This section summarizes some of these benefits, with specific examples.

Cloning to Improve Understanding of Human Disease

The creation of cloned embryos using nuclei from individuals carrying genetic mutations – specifically, genes that predispose them to particular diseases – might be used to better understand and treat those diseases. Consider, for example, Parkinson's disease. A characteristic of Parkinson's disease is the aggregation in dying brain cells of a protein called alpha-synuclein. Two different mutations in the alpha-synuclein gene produce forms of the protein that aggregate more readily. Individuals carrying these gene mutations suffer from early-onset Parkinson's disease.

To study how genetic disease develops, scientists look for suitable laboratory models. One strategy for producing such disease models is to inject the disease-causing human genes into human or animal cells in tissue culture to produce a cell-system expressing the abnormality. Although it has been possible to introduce copies of mutant genes into various kinds of human and animal cells, the resulting in vitro cell-systems imperfectly model the human disease. In part this is because the behavior of specific proteins within cells is influenced by their interactions with other cellular proteins. For example, human alpha-synuclein in a mouse cell cytoplasm interacting with mouse proteins is unlikely to behave the same way that it does in a human cell surrounded by human proteins. To study human disease, it is generally preferable to work with human cells and tissues.

A preferable alternative to introducing mutant genes into normal cells is to begin with human cells that are already abnormal – in this case, cells carrying the mutant genes that predispose their bearers to Parkinson's disease. If one could obtain embryonic stem cells derived from cloned embryos produced using nuclei from individuals carrying these mutant genes, one could then stimulate them to differentiate into dopamine – producing nerve cells in vitro. These cells would provide a vastly improved model for understanding the metabolism of alpha-synuclein and its role in the development of Parkinson's disease.< In this example, the availability of improved in vitro models for genetic and neurodegenerative diseases could shorten the time required to understand them and to devise new treatments.

It is true that adult stem cells (or multipotent adult progenitor cells), isolated from patients carrying the mutant genes that predispose them to Parkinson's disease, might also be stimulated to become dopamine-producing neurons in vitro. But there are unanswered questions about the ease of culture and long-term viability of such cells, and the likelihood of success with cellular models of disease derived from adult stem cells remains unknown. In the absence of a certain and superior alternative, it would be wrong to forgo the possibly unique benefits of cloning for disease research.

Cloning to Devise New Treatments for Human Diseases

The same cellular model systems used to study disease processes are also potentially useful for assessing and developing chemical or pharmaceutical treatments for the disease in question. To continue with the Parkinson's disease example, neurons derived from stem cells containing the alpha-synuclein aggregation mutations would be very useful for testing compounds that might prevent aggregation of this protein. Chemicals that effectively prevented aggregation in this model system could be useful starting points for the development of new drugs for the specific treatment of Parkinson's disease. Here, too, neuronal cell-systems derived from adult stem cells carrying the mutations might serve as well as those derived from cloned embryonic stem cells. But there is no way of knowing in advance which of the alternative routes is more promising. From a medical and scientific point of view, research on cloned embryos may offer unique benefits.

Cloning to Produce Immune-Compatible Tissues for Transplantation

Some animal studies suggest that tissues derived from embryonic stem cells can, if injected under certain conditions, populate disease-stricken areas and differentiate so as to compensate for the loss of function caused by the diseased tissue. For example, liver or heart muscle cells injected into an animal with liver or heart disease could help regenerate the diseased tissues and restore normal function. But these cells would have a chance to do this only if they can survive the normal immunological rejection response to foreign material. Cloning-for-biomedical-research offers the possibility that scientists could someday generate individualized, "rejection-proof" replacement cells and tissues to help patients fight disease and restore health. Stem cells and tissues derived from an embryonic clone of the patient would have the same genes as the patient, and so, hypothetically, would not be rejected by the patient's body as foreign.

It is true that this possibility (what is sometimes called "therapeutic cloning") remains unproved. As before, there may be alternative (nonembryonic or adult) sources of such "rejection-proof" stem cells and tissues derived from them. And there is ongoing research to circumvent the rejection problem altogether, by, for example, modifying the surface of an unrelated (embryonic) stem cell so as to enable it to escape detection as "foreign" tissue when transferred to patients for therapy. But, once again, it is too early to say which approach will work, and therefore it is important, from a medical and scientific perspective, not to close off any avenue of promise. The only way to verify this hypothesis is to try it – first in animals, then in human volunteers.

Cloning to Assist in Gene Therapy

Cloning techniques could also be combined with precise genetic manipulation to devise genetic treatments for genetic diseases. For example, a cloned embryo produced from a patient with severe combined immunodeficiency could be genetically modified to correct and repair the disease-causing mutation. Stem cells taken from the genetically modified cloned embryo might then be used to develop bone marrow stem cells to transplant back into the patient. This combined approach to gene therapy has shown early promise in one attempt to correct a genetic abnormality in the immune system of mice.

Possible Moral Dilemmas of Proceeding

The potentially unique medical benefits of cloning-for-biomedical-research are, to those of us who favor it, abundantly clear. Yet the moral meaning of proceeding, still to be considered, is the subject of some debate among us. Most of us who favor proceeding believe that this area of promising research is nonetheless fraught with moral quandaries and ethical trade-offs; a minority of us do not share these concerns. The minority view, labeled Position Number Two, follows the principal moral case for cloning-for-biomedical-research under strict limits, designated here as Position Number One. Each opinion is presented in turn.

Position Number One

What makes this research morally controversial is that it involves the production, use, and intentional destruction of cloned human embryos. To determine whether or not the science should proceed – or, if it does, what limits should be placed on this research – it must be asked what, if anything, is owed this nascent form of human life. Only then can an evaluation be made of whether the possible benefits of this research justify its potential human cost. Other moral hazards must be considered that are either inherent in, or possible consequences of, this line of research. These hazards include the following: the possibility that cloned embryos will be developed and experimented upon beyond the blastocyst stage (the stage from which stem cells are taken); the possible exploitation of women who would be donors of eggs; the possibility that the production of cloned human embryos will lead

– intentionally or unintentionally – to cloning-to-produce-children; and the possibility that engaging in such research will weaken or undermine society's respect for human life, and therefore undermine the very good (life) that it is meant to serve. Each of these moral challenges will now be addressed.

(a) What is owed to the cloned embryo? The subject of the moral status of developing human life is a difficult and controversial matter, one about which American society is and appears likely to remain deeply divided.

We are well aware of the fact that we cannot do it full justice in the present context. Yet we believe that the moral defense of cloning-for-biomedical-research requires a consideration of what is owed nascent human life (cloned or not). There is also the question – considered at great length in Chapter Three – of whether cloned embryos are the moral equivalent of fertilized embryos, or whether the different nature of their origins and the uncertainty of their capacity to become full human beings means that our moral duties to them are somehow different.

Nevertheless, those who wish to defend cloning-for-biomedical-research-as we do here – must consider what is owed to *embryos as such* as well as the significance of the fact that the embryos in question would be **cloned**. That said, the relevant arguments, especially in this subsection and the next, are in most crucial respects the same as those regarding the treatment of embryos produced by IVF.

Let us be clear about what we are talking about when we speak of cloned *embryos*. We are talking about the very earliest stages in development, from the single cell product of SCNT, through the early cleavage stages, up to the blastocyst stage. This is a structure comprising some 100 to 200 cells not yet differentiated into specific tissues, let alone organs (though there is differentiation into inner cell mass and trophoblast; see Chapter Four). It is true that the embryos at the blastocyst stage, if implanted in a woman's uterus or (hypothetically) an animal or artificial womb, could be made to develop to later stages, and this potentiality must be taken into account. But it is important to keep in mind the primitive and undifferentiated condition of the embryonic stage that is relevant for the research in question.

We begin with a series of questions: Is destroying an embryo or cloned embryo at the blastocyst stage morally the same as killing a child? Is it the same as clipping a fingernail? Is it more like one of these acts than the other? Is it like neither? Does the moral status of an embryo depend on whether it is implanted in a woman's uterus or remains in a laboratory? Does the moral status of an embryo depend on its origins, or how it was produced? Does it depend on the motives of those who create it?

In our view, embryos have a developing and intermediate moral worth, such that the early human embryo has a moral status somewhere between that of ordinary human cells and that of a full human person. We acknowledge the difficulty of setting perfectly clear lines marking when an embryo's moral status goes from "less than a human person" to "like a human person" to "fully a human person." But we believe there are sound moral reasons for not regarding the embryo in its earliest stages (certainly in the first fourteen days) as the moral equivalent of a human person, though it does command significantly more respect than other human cells. We also hold that the embryo can be used for life-saving or potentially life-saving research while still being accorded the "special respect" it deserves, and while still preventing abuses such as research on later-stage embryos or fetuses or the production of cloned children. We will develop this view by taking up the significance of (i) twinning, (ii) implantation, (iii) the human form, and (iv) the notion of "special respect."

(i) The possibility of twinning. First, it is still unclear in the initial fourteen-day period whether an embryo will develop into one or more human beings. The possibility for "twinning" is still present, suggesting that the earliest-stage embryo is either *not yet* an individual or is a being that is not confined to becoming *only one* individual. There are continuing philosophical debates about how to understand what happens in twinning: for example, whether one individual embryo "clones" itself to produce a second, or whether an organism that resembles (but is not *yet*) an individual embryo divides into two truly individual beings. Nevertheless, the biological – and we believe moral – significance of the possibility for twinning is clear: after fourteen days (or after the primitive streak is formed), the being in question *can no longer be anything*

but a single being – that is to say, no embryo after this stage, and thus no fetus or live-born baby, can replicate or divide to form another identical being. Before fourteen days, this possibility remains.

(ii) The moral significance of pregnancy and implantation. Both IVF embryos and cloned embryos in vitro differ from comparable embryos conceived through sexual intercourse, for two reasons. First, the possibility for pregnancy with IVF or cloned embryos requires human assistance – that is, it requires the medical procedure of transferring an embryo into the woman's uterus. There is thus no possibility of the IVF or cloned embryo becoming a human child in its original in vitro environment. Second, embryos that are conceived through sexual intercourse have a direct physical connection with the individual women who carry them, whereas an in vitro embryo (cloned or not) has no such connection unless it is transferred into a woman's uterus. Thus, transfer of cloned or IVF embryos into a woman's uterus is a significant moral step, insofar as such embryos cannot be removed – they can never again be held in human hands – without a direct physical intrusion or violation of the pregnant woman. Of course, it might become technologically possible in the future for in vitro embryos to develop beyond the blastocyst stage – and perhaps even to birth – without implantation into a woman's uterus (that is, in an artificial womb). Moreover, just because those embryos (cloned or not) that exist in vitro cannot continue to develop in a self-directed way beyond the blastocyst stage – that is, they require human artifice of some kind to develop further – does not mean that the preimplantation embryo is morally insignificant. But implantation does mark a significant point in these two respects: after implantation, self-direction toward birth (without external human artifice) becomes *possible* and external human control of embryos becomes *impossible* without intruding upon or violating the pregnant woman.

(iii) The significance of the developed human form. Generally speaking, our moral sentiments respond very differently to the prospect – or the sight – of the destruction of an embryo and the murder of a child. In other words, there is a difference between *what we respect and what we consider inviolable.* The destruction of

embryos might inspire concern or solemnity. In contrast, our reaction to the murder of a child would be one of horror, outrage, grief, and violation. James Q. Wilson has discussed how these two fundamentally different moral reactions change as the embryo develops into a fetus and then into a child – and correspondingly, how our concern and solemnity transform into horror and outrage. Specifically, human beings exhibit a distinctly different moral sympathy for, and therefore greater willingness to protect, those organisms that have begun to resemble human beings in their developed form. The practice of sacrificing the life of the unborn in order to save the life of the pregnant woman – while not a moral parallel to the case of using cloned embryos for biomedical research – shows that there is some moral precedent for subordinating nascent human life to more developed human life. Of course, taken to an extreme, such a principle would justify the most grotesque uses of developing human fetuses for scientific experiments. Moreover, the case is not strictly analogous, for in the case of the pregnant woman, two lives are in conflict, a confrontation absent with free-standing embryos. We do not take the life of woman A's unborn fetus to save the life of woman B, not even with consent. But these difficulties notwithstanding, there is (again) a moral insight in this example. It demonstrates the important moral obligation of caring for those who already dwell among us, and the inevitable moral complexity of weighing different forms of human life, especially nascent and developed human life, against one another. It also suggests ways in which the claim on our protection may increase with the emergence of powers of awareness and suffering. Of course, such examples – and our moral sentiments in general – are not by themselves decisive. They are the beginning, not the end, of reasoning about our moral responsibilities. But they should also not be ignored for what they reveal about the nature of particular beings and particular acts – and in this case, for what they suggest about both the *developing* and *intermediate* status of the early human embryo.

(iv) The meaning of "special respect." Finally, there is the question of whether it is possible to accord early-stage embryos "special respect" while still using them for biomedical research. We might reason here by an admittedly imperfect analogy. Various religions

have rules governing the killing of animals for food. These exist in part to restrain cruelty. But they also serve to demonstrate respect for beings that command our affections and our wonder, because they are (like us) part of the mystery of existence. In a similar way, many hunters have a deep-rooted respect and even affection for the animals they kill. This is not to say that human embryos are the same as animals, because, in our opinion, they are indeed human organisms, if not fully developed human beings. But it is to show that there might be ways both to respect beings and to use them – for serious, not frivolous, reasons, and as part of our place in the order of being, not simply as an extension of our subjective will.

For the above-stated reasons, we would assign an intermediate and developing status to the human embryo. Those who treat the developing early embryo as nothing more than "mere cells" (see Position Number Two below) are in danger of ignoring its direct and inherent connection to the profound mystery of the origins of human life and seem willing to ignore the fact that an embryo will (and a cloned embryo might) eventually become one (or more) human being(s). This view greatly underestimates the moral seriousness of the question of whether to proceed with research on nascent human life. And it gravely mischaracterizes the meaning of potentiality – specifically, the difference between having the capacity to become anything at all (a pile of building materials, for example) and the capacity to become something in particular (an individuated human person or persons).

At the same time, those who believe that early-stage embryos are the moral equivalent of a human person (see Part IV below) are also, we believe, misguided. Just as we must listen to – and then articulate – the moral meaning of our disquiet at the idea of cloning-to-produce-children, we must listen to and articulate our fundamentally different moral responses to the destruction of an embryo on the one hand and the murder of a child on the other. While no single criterion like "appearance," "self-consciousness," "the capacity to express needs and desires," or "the capacity to feel pain" can by itself be decisive in conferring human dignity, the absence of all such criteria in the early-stage embryo or cloned embryo suggests that it is not a truly human being, but something different, commanding our respect because of what it is and may become, but yet not fully one of us.

In sum, what is owed the embryo is not the same protections, attachments, and rights as a human person; nor is it no respect at all. In making the decision to proceed with research on embryos or cloned embryos, we must do so only for the most compelling reasons – namely, the reasonable expectation that such research will save human lives – and only with eyes open to the moral burden of doing what we believe to be morally best. Even as we establish the biological and moral grounds for using human embryos in certain forms of research, we must face and accept the solemnity of what we propose. Finally, we must proceed with the paradox that accompanies all human suffering and human imperfection in full view: that sometimes we seem morally obligated to do morally troubling things, and that sometimes doing what is good means living with a heavy heart in doing it.

(b) The problem of deliberate creation for use in research. We next address whether the creation of embryos explicitly for the purposes of biomedical research presents additional ethical problems, beyond those just examined. In the case of research on cloned embryos, this form of deliberate production and destruction – rather than the use of leftover embryos initially created for reproductive purposes – is the only means of proceeding, if, at the same time, society prohibits cloning-to-produce-children. It is one thing to overcome the respect owed to an already existing embryo that would die even if not used for research. It is, some argue, quite another thing to bring the embryo into being solely for use and exploitation in research. Willing to accept the first, they reject the second. In this connection, three issues seem worth considering.

First, the fundamental moral judgment about whether to proceed with cloning-for-biomedical-research must be grounded in our judgment about the moral status of the embryos themselves, not the purpose of their creation. If an embryo or a cloned embryo had no moral standing, then creation for research and eventual destruction would present no moral problem. If the embryo or cloned embryo were morally the equivalent of a child, then regardless of how or why it was produced, experiments upon it would be morally abhorrent. But if, as we have just argued, an embryo or a cloned embryo has a developing and intermediate moral status, certain worthy uses of them may be justified regardless of how and why they were produced. Because the use of stem cells from cloned embryos may in the future provide treatment for serious human diseases, the creation of cloned embryos and their subsequent disaggregation to isolate stem cells can be justified.

Second, the moral responsibilities for producing new embryos *solely* for research and for producing extra IVF embryos *later used* in research are not really so different. In the case of IVF and leftover embryos, the individuals who create them for reproductive purposes typically and deliberately create more embryos than they are likely to use, and therefore know in advance that some will probably be destroyed. It is true that they are produced with the intent of initiating a pregnancy and that the embryo wastage is not all that different from what obtains in efforts to conceive in vivo. But the moral responsibility for production, use, and destruction of leftover embryos are finally no less than for deliberate production for use (and subsequent destruction in research). (We acknowledge that some who accept this logic come to the opposite conclusion – namely, not that cloning-for-biomedical-research is morally permissible but that IVF should be morally restricted to creating one embryo at a time, if permitted at all.)

Third, in both cases – creating embryos to aid fertility or creating embryos for biomedical research – the ultimate goal is something humanly good: a child for an infertile couple or research that holds promise for curing debilitating diseases and easing suffering. Thus, in the case of cloning-for-biomedical-research, it is wrong to argue, as some do, that embryos are being "created for destruction." Certainly, their destruction is a known and unavoidable effect, but the embryos are ultimately created for research in the service of life and medicine.

In the end, while we acknowledge the risk of turning nascent human life into a "resource" – fully separate from its intrinsic connection to human procreation – we hold that the concern over deliberate creation and destruction is misplaced. What matters instead is whether a proper regard is shown for the created embryos, and therefore whether a proper moral and legal framework can be established that limits and governs their use in accordance with the respect they are owed as *human* cloned embryos.

(c) Development and use of cloned embryos beyond the earliest stages. A perceived danger of allowing cloning-for-biomedical-research is that some researchers will develop cloned embryos beyond the blastocyst stage for research purposes. There are good scientific reasons and even moral arguments for doing so: one could learn much more about development, normal and abnormal, by going to later stages; and differentiated tissues taken from cloned fetuses would likely be more useful in regenerative medicine than stem cells. There is already at least one animal study showing the potential of this approach. Transplantable functioning kidney tissue has been attained from six-week-old cloned cow fetuses, developed from cloned cow embryos transferred into a cow's uterus for partial gestation. Cloned human embryos might be developed past the blastocyst stage by implantation into an animal or human uterus, by the development of artificial wombs, or by advances in sustaining nascent human life in vitro.

This is a serious concern for those of us who believe that the cloned embryo has only an intermediate moral status and who also recognize the difficulty of drawing bright lines for when developing human life changes from "less than a human person" to "like a human person" to a "fully developed person." Clearly, the longer cloned embryos are allowed to develop, the more severe the moral burden in using them. And at some point, the moral burden of proceeding becomes a moral obligation not to proceed – even if significant medical benefits might be gained from doing so. In such circumstances, the medical principle of "do no harm" must override the researcher's desire to do good, lest we undermine the humanistic principles and spirit of the entire medical enterprise.

The moral tradition of "erecting a fence around the law" may provide a useful guide in this case. We recommend that research on cloned embryos be strictly limited to the first fourteen days of development – a point just about when the primitive streak is formed and before organ differentiation occurs. We acknowledge that by erecting the fence more widely, we might be more certain to prevent this particular abuse (developing cloned embryos beyond the blastocyst stage). We also acknowledge that relaxing this limit to permit research beyond fourteen days might yield additional medical benefits. There is a moral burden in both directions. But we hold that there is a point of development beyond which research on nascent human life is morally intolerable no matter what the potential medical benefits. By raising a permanent fence at fourteen days, the dignity of human life will be sufficiently protected.

(d) Exploitation of women who are egg donors. Additional concerns in proceeding with cloning-for-biomedical-research are the possible dangers to, and exploitation of, women who are egg donors. The removal of eggs remains an unpleasant and (owing to the hormone treatments needed to hyperstimulate the ovaries) a risky medical procedure for women. It is therefore restricted mostly to circumstances where such a procedure is necessary to treat infertility – that is, where the women themselves are the beneficiaries of the procedure. Moreover, one possible avenue of cloning-for-biomedical-research-namely, the creation and future use of individualized stem cells – would potentially require, if it became feasible, a very large and indefinite number of eggs.

These are genuine concerns. But they can be addressed by strictly adhering to the established body of ethics for research on human subjects. These ethical codes suggest the following requirements: regulation to prevent the creation of improper financial incentives for participating in such research; full disclosure by the users of human eggs of their practices; a commitment to consider using nonhuman eggs, so as to decrease the need for human egg donors ; and strict limits on the uses of cloned embryos for only those investigations that uniquely require them.

(e) The connection to cloning-to-produce-children. The final moral concern is that cloning-for-biomedical-research will lead – intentionally or not – to cloning-to-produce-children. For the reasons described in Chapter Five, we believe that the creation of cloned human children would be unethical and that society has a moral responsibility to ensure that this does not happen. Thus we are obliged to consider whether the pursuit of cloning-for-

biomedical-research is consistent with a serious commitment to stopping cloning-to-produce-children. A number of points must be considered.

First, the production of cloned embryos, even for research purposes, crosses a new line by bringing into existence for the first time forms of nascent human life that are asexually produced. Second, experience with producing cloned embryos for biomedical research might well improve the technique of cloning itself, and therefore result in the greater perfection of the first step toward cloning-to-produce-children. Third, cloning-for-biomedical-research means that cloned embryos would exist in laboratories where they could be available for efforts to initiate a pregnancy. Finally, a society that allows cloning-for-biomedical-research, while setting strict legal limits on cloning-to-produce-children, will likely require the mandatory destruction of nascent human life.

The first concern is intrinsic to cloning-for-biomedical-research in itself. Are we a different society because we have brought asexually produced human embryos into existence? In some ways, perhaps we are. We are confronted by the scope of our powers to change human life, to alter human procreation, and to modify the nature of human origins and the genetic makeup of new life. But we are also reminded of what should be the animating purpose of that power: to cure disease and relieve suffering. We are reminded of both new and unique possibilities for human harm (from the production of human clones) and new and unique possibilities for human benefit (from research on cloned embryos). This is, we suggest, the meaning of crossing this line.

The second and third concerns are connected to where this research might lead: namely, to a perfected cloning technique and to the intentional production of cloned children. This is indeed a genuine concern. It is perhaps the case that the best way to prevent the production of cloned children is to prohibit the creation of cloned embryos. But in the end, we are not convinced that cloning-for-biomedical-research will inevitably lead to cloning-to-produce-children; rather, we believe that the best approach is a system of regulation that prevents such an abuse. Such a system would include: a legal ban on the implantation of cloned embryos in *any* uterus (human, animal, or artificial); a prohibition on developing cloned embryos beyond fourteen days; a requirement that any individual or group engaging in cloning-for-biomedical-research register with proper regulatory authorities; prior scientific review of all proposed uses of cloned embryos to judge their medical and scientific benefits; and strict accounting of all cloned embryos that are produced to prevent their removal from the lab of origin or their use in attempts at cloning-to-produce-children.

Of course, no system of regulation is perfect. There is always the possibility of malfeasance or error. The prudential question in this case is whether the likelihood of cloning-to-produce-children is increased – at all, slightly, or significantly – by allowing the production and use of cloned embryos for biomedical research. But there is also the question of whether some additional risk of cloning-to-produce-children is justified or tolerable given the human goods that might be achieved through cloning-for-biomedical-research. In our view, it is.

The final concern is that to pursue research on cloned embryos while preventing cloning-to-produce-children would require laws that mandated the destruction of nascent human life. In assessing the moral significance of this fact, we return to our judgment about the moral status of cloned embryos, what is owed to them, and whether the human goods that can be achieved by cloning-for-biomedical-research justify the real and potential human costs. In our view, the possible existence of a law requiring the destruction of cloned embryos at or before fourteen days of development would force moral clarity about what we are doing – and the burdens of doing it. Such a law might remind society of the ambiguity and limits of the efforts to "heal the world," and therefore the dangers of trying to do so by any means possible. The need for such a law requiring the destruction of nascent human life would also remind us that there is a burden in acting just as there is a burden in not acting.

(f) Conclusion. The case for cloning-for-biomedical-research – as with all research that involves the use of nascent human life– should not consist simply of guessing how many people might be saved and how many embryos might be lost. The moral concerns cannot so simply be taken up, addressed, and retired. They are permanent concerns and permanent burdens.

We believe, in this particular case, that the promise of cloning-for-biomedical-research justifies proceeding, but that the genuine possibility of moral harm requires strict regulations of how we proceed. We have tried to articulate what such a system of regulation might include: (1) a legal requirement not to develop cloned embryos beyond fourteen days of development and not to implant cloned embryos in any uterus, human, animal, or artificial; (2) the creation of a governmental oversight body to regulate individuals and groups who engage in this research, and to account for all cloned embryos that are produced so as to prevent their removal from the lab of origin or their use in cloning-to-produce-children; (3) a ban on commerce in living cloned human embryos; (4) adherence to the highest standards of the ethics of research on human subjects, especially when it comes to procuring eggs; (5) a prior scientific review of the proposed uses of cloned embryos to judge their unique medical and scientific benefits; and (6) continued research into possible non-embryonic sources of stem cells and tissues for developmental studies, and ways other than cloning to solve the immune rejection problem. Such regulations amount to much more than mere bureaucratic red tape. They embody a profound ethical insight – namely, that the means of serving human beings must never corrupt our responsibilities to human beings.

Position Number Two

A few of us who favor proceeding with cloning-for-biomedical-research have few of the ethical qualms expressed by our colleagues in Position Number One. It is our view that this research, at least in the forms and for the purposes presently contemplated, presents no special moral problems, and therefore should be endorsed with enthusiasm as a potential new means of gaining knowledge to serve humankind. Because we accord no special moral status to the early-stage cloned embryo, we believe that the moral issues involved in this research are no different from those that accompany many existing forms of biomedical research, requiring mainly the usual commitment to high standards for the quality of research, scientific integrity, and the need to obtain informed consent from, and to protect the health of, donors of the eggs and somatic cells used in nuclear transfer.

It is also our view that there are no sound reasons for treating the early-stage human embryo or cloned human embryo as anything special, or as having moral status greater than human somatic cells in tissue culture. A blastocyst (cloned or not), because it lacks any trace of a nervous system, has no capacity for suffering or conscious experience in any form – the special properties that, in our view, spell the difference between biological tissue and a human life worthy of respect and rights. Additional biological facts suggest that a blastocyst should not be identified with a unique individual person, even if the argument that it lacks sentience is set aside. A single blastocyst may, until the primitive streak is formed at around fourteen days, split into twins; conversely, two blastocysts may fuse to form a single (chimeric) organism. Moreover, most early-stage embryos that are produced naturally (that is, through the union of egg and sperm resulting from sexual intercourse) fail to implant and are therefore wasted or destroyed.

There is a moral precedent for using materials from early human embryos in the widely accepted practice of using organs from brain-dead human beings. Upon determination of death, and with permission from the next of kin, surgeons routinely harvest organs to save the lives of sick or dying patients. In a similar way, donors of somatic cells and human oocytes could justifiably grant a biomedical scientist permission to use cells derived from the resulting cloned five-to-six-day-old blastocyst, which also completely lacks a brain and a capacity for consciousness.

Some argue that the transplantation analogy is misleading, because a blastocyst has the potential to become a fetus and ultimately a child, whereas the brain-dead individual does not. But the *potential* to become something (or someone) is hardly the same as *being* something (or someone), any more than a pile of building materials is the same as a house. A cloned embryo's potential to become a human person can be realized, if at all, only by the further human act of implanting the cloned blastocyst into the uterus of a woman. Such implantation is not a part of cloning-for-biomedical-research, whose aims and actual practice do not require it.

Moreover, thanks to the results of nuclear transplantation research, there is reason to believe that every human cell has the genetic potential to develop into a complete human being, if used in cloning efforts to produce a child. If mere potentiality to develop into a human being is enough to make something morally human, then every human cell has a special or inviolable moral status, a view that is patently absurd.

"Slippery slope" warnings that the use of early-stage cloned embryos for research would lead necessarily either to the production of cloned children or to research on later-stage cloned fetuses should be treated with skepticism. Appropriate regulations can easily be established and enforced to prevent any such abuses. Although the continuity of biological development means that there is no naturally given moment after which an embryo or fetus becomes a person, defensible boundaries can be set. It is perfectly possible to treat a blastocyst as a clump of cells usable for lifesaving research, while prohibiting any such use of a later-stage embryo or fetus.

Where to set the boundary is a matter for prudent judgment. For the foreseeable future, the moral line might be safely drawn at fourteen days of development, when no nervous system has developed and when a distinct identity as a single individual has not yet been preordained. Also, derivation of the valuable stem cells can be accomplished well before fourteen days. Whether society will be faced, in the future, with reason to reconsider such a line is for now a matter of speculation. If such an occasion ever arose, it would require an evaluation of the proposed scientific use and its likely medical benefits and a moral consideration of whether the research in question justified using embryos beyond the fourteen-day point.

Bibliography

Acquaah, George: *Understanding Biotechnology: An Integrated and Cyber-Based Approach*, NJ, Pearson/Prentice Hall, 2004.

Ausubel, F.M.: *Current Protocols in Molecular Biology*, New York, John Wiley and Sons, 1989.

Baker, Kenneth Frank: *Ecology of Soil-bourne Plant Pathogens, Prelude to Biological Control*, Berkeley, University of California Press, 1965.

Bauer MW: *Biotechnology-the Making of a Global Controversy*, Cambridge, Cambridge University Press, 2002.

Bernauer, Thomas: *Genes, Trade, and Regulation: The Seeds of Conflict in Food Biotechnology*, Princeton, Princeton University. 2003.

Broach, J.R.: *The Molecular Biology of the Yeast*, Cold Spring Harbor, Cold Spring Harbor Laboratory, 1981.

Chrispeels, Maarten : *Plants, Genes and Crop Biotechnology*, Sudbury MA, Jones and Barlett Publishers, 2003.

Costanza, R.: *Ecological Economics*, Columbia Uni. Press., New York, 1991.

Daubenmire, F.: *Plants and Environment*, New York, Wiley, 1947.

Dusenbery, David B.: *Life at a Small Scale: the Behavior of Microbes*, New York, Scientific American Library, 1996.

E. Dereniak and D. Crowe, *Optical Radiation Detectord*, John Wiley and Sons, 1984.

F. Reif: *Fundamentals of Statistical Mechanics*, McGraw-Hill, Singapore, 1965.

Fransman M, Junne G, Roobeek A: *The Biotechnology Revolution?*, Oxford, Blackwell, 1995.

Fumento, Michael: *Bioevolution: How Biotechnology is Changing Our World*, San Francisco, Encounter Books, 2003.

Goodman, D.C.: *From Farming to Biotechnology: A Theory of Agro-industrial Development*, Oxford, Blackwell, 1987.

Jameson, J. L.: *Principles of Molecular Medicine*, Totowa, Humana Press, 1998.

Li, Koibong and Leslie A. Lewis.: *A Novel Approach to the Evolution of Microorganisms*, New York, Vantage Press, 1987.

Lodish, H.: *Molecular Cell Biology*, New York, Freeman, 2000.

Lynch, J.M., and Hobbie, J.E.: *Micro-organisms in Action: Concepts and Applications in Microbial Ecology*, Blackwell Scientific Publications, Oxford, 1988.

Maarten, C.: *Plants, Genes and Crop Biotechnology*, Sudbury MA, Jones and Barlett Publishers, 2003.

Metting, Jr., F.B.: *Soil Microbial Ecology: Applications in Agricultural and Environmental Management*, Marcel Dekker, New York, 1992.

Raff, K. Roberts, and J. Watson, *Molecular Biology of the Cell*, New York: Garland Publishing, 1989.

Rietman, Ed.: *Molecular Engineering of Nanosystems*, New York, Springer, 2001.

Sayler, G.S., Fox, R., and Blackburn, J.W.: *Environmental Biotechnology for Waste Treatment*, Plenum Press, New York, 1991.

Sheldon, J.K.: *Practical Environmental Bioremediation*, Lewis Publishers, Boca Raton, 1992.

Switzer, R.L.: *Experimental Biochemistry*, New York, W.H. Freeman and Company, 1977.

Teich, M.: *Fundamentals of Photonics*, New York: Wiley, 1991.

Wood, E.J. Ferguson: *Marine Microbial Ecology*, London, Chapman and Hall, 1965.

Yanoff, M.; Duker, J. S.: *Ophthalmology*, St. Louis, Mosby, 2004.

Index

A

Applications, 32, 37, 72.
Appropriate Terminology, 184.
Asexual Reproduction, 137, 146, 158, 181, 184, 196, 197, 203.

B

Biomedical Ethics, 2, 3, 26, 31, 37, 40, 45, 46, 111, 120.
Biomedical Research, 10, 28, 37, 38, 51.
Biomedical Research, 3, 10, 11, 28, 31, 34, 36, 37, 38, 42, 51, 53, 57, 59, 60, 61, 117, 135, 136, 137, 139, 144, 147, 148, 149, 150, 152, 153, 156, 162, 163, 178, 179, 182, 195, 245, 246, 247, 248, 258, 260, 261, 264, 265, 266.
Biotechnology, 156, 160, 177, 206.

C

Clinical Research, 3, 13, 14, 21, 27, 28, 31, 37, 38, 61, 70, 71, 72, 73, 74, 82, 90, 93, 99, 105, 106, 109, 111, 117, 126.
Collaboration, 37, 42.
Continuity, 24, 31, 73, 92, 93, 94, 95, 97, 107, 191, 268.
Contribution, 7, 17, 25, 108, 180, 244.

D

Delivery Research, 3, 23, 24, 26, 27, 29, 31, 32, 34, 36, 41, 46, 50, 90, 94, 108, 127, 129.
Development, 5, 6, 11, 12, 16, 18, 22, 25, 37, 38, 39, 42, 44, 49, 51, 53, 56, 57, 58, 59, 60, 66, 67, 69, 70, 72, 98, 99, 100, 107, 108, 122, 123, 136, 137, 138, 142, 144, 155, 157, 162, 164, 168, 172, 178, 180, 183, 185, 187, 188, 189, 190, 191, 192, 193, 194, 195, 197, 198, 199, 200, 202, 203, 219, 223, 233, 240, 242, 251, 252, 253, 255, 262, 263, 265, 266, 268.
Disease Prevention, 14, 24, 25, 31, 50, 63, 64, 93, 105, 107, 108, 109.

E

Ecology, 195.
Evidence, 1, 19, 30, 56, 59, 70, 78, 84, 96, 100, 101, 102, 105, 123, 148, 154, 199, 220, 228.
Evolution, 37, 170.

G

Government, 3, 23, 30, 39, 40, 42, 46, 88, 100, 120, 121,

129, 146, 150, 151, 176, 233, 239.

H

Health Promotion, 14, 24, 25, 31, 50, 63, 93, 105, 107, 108, 109.
Health Services, 2, 3, 14, 15, 23, 24, 25, 26, 27, 29, 31, 32, 34, 36, 37, 43, 45, 46, 49, 50, 61, 69, 90, 92, 93, 94, 99, 100, 101, 104, 105, 106, 107, 108, 109, 115, 117, 122, 126, 127, 129, 131, 132.
Human Procreation, 141, 156, 157, 160, 165, 170, 206, 211, 212, 214, 224, 225, 226, 227, 230, 232, 233, 239, 261, 264.

I

Institutions, 3, 24, 35, 37, 38, 39, 42, 57, 62, 107, 112, 124, 125, 127, 160, 165.
Investment, 29, 35.

M

Management, 14, 21, 25, 43, 61, 62, 65, 66, 68, 81, 89, 97, 108, 116, 122, 123.
Mental Health Services, 24, 25, 31, 93, 104, 105, 107, 108.

N

National Research Agenda, 1, 4, 10, 30, 31, 46, 47, 61.

O

Opportunity, 1, 2, 19, 22, 45, 48, 62, 65, 71, 83, 112, 113, 237.

P

Population Ecology, 195.
Population Trends, 43, 90.
Products, 18, 70, 141, 157, 164, 178, 180, 192, 224, 226, 227, 230, 232, 233.
Protection, 118, 119, 135, 150, 164, 185, 215, 216, 219, 231, 247, 258.
Proteins, 11, 12, 54, 56, 169, 206, 210, 260.
Public Policy, 47, 81, 135, 146, 147, 153, 163, 170, 177, 239, 242.

R

Relationship, 8, 12, 20, 24, 63, 77, 81, 96, 99, 106, 109, 126, 135, 165, 206, 227.
Religious, 176, 181.
Research Priorities, 4, 7, 12, 20, 31, 35, 53, 54, 61, 62, 79, 80, 93, 106, 108, 109, 111, 117.

S

Security, 92, 98, 104.
Social Research, 19, 22, 23, 24, 28, 29, 31, 34, 73, 75, 76, 86, 88, 132.

T

Technology, 6, 13, 22, 24, 38, 39, 53, 58, 59, 86, 99, 107, 112, 117, 121, 133, 135, 146, 149, 152, 153, 155, 156, 160, 161, 162, 163, 164, 165, 179, 203, 207, 210, 223, 226, 231, 234, 235, 243.
Terminology, 136, 180, 183, 184, 190.

□□□